Self-Organization in Embedded Real-Time Systems

M. Teresa Higuera-Toledano • Uwe Brinkschulte
Achim Rettberg
Editors

Self-Organization in Embedded Real-Time Systems

Springer

Editors
M. Teresa Higuera-Toledano
Facultad de Informática, DACYA
Universidad Complutense de Madrid
Calle del Profesor Gaecía
Santesmas
Madrid, Spain

Uwe Brinkschulte
Fachbereich Informatik und Mathematik
Goethe University
Robert Mayer-Straße 11-15
Frankfurt am Main
Germany

Achim Rettberg
OFFIS e.V
Carl von Ossietzky University
Oldenburg
Escherweg 2
Oldenburg, Germany

ISBN 978-1-4614-1968-6 ISBN 978-1-4614-1969-3 (eBook)
DOI 10.1007/978-1-4614-1969-3
Springer New York Heidelberg Dordrecht London

Library of Congress Control Number: 2012948873

Printed on acid-free paper

Springer is part of Springer Science+Business Media (www.springer.com)

Preface

Self-organization is a key technique to handle the growing complexity of upcoming computing systems. This refers to both hardware and software. Distributed systems running hundreds of tasks on dozens of processors, each equipped with multiple cores, require self-organization principles to ensure efficient and reliable operation. Therefore, substantial research has been done in the last years to address the so-called Self-X features like self-configuration, self-optimization, self-adaptation, self-healing, self-protection, etc. Additionally, real-time properties are another key issue in many complex systems (e.g., in the embedded systems area).

Over the last 5 years, the supporting Self-X-related technologies, in this increasingly important area, have been developed. The research on self-organization real-time technology has been established during these last years and is producing an increasing number of mature results regarding both hardware and software in several areas as: *(i)* Open components for embedded real-time adaptive and self-organizing applications. *(ii)* Innovative techniques in scheduling, quality of service, communications supporting organic real-time applications. *(iii)* Multi-/Many-Core embedded systems supporting real-time adaptive systems. *(iv)* Hardware support for operating systems executing real-time organic applications. *(v)* Real-time communication systems for distributed real-time adaptive applications. *(vi)* Power-aware adaptive hardware and software systems. *(vii)* Case studies of open embedded real-time self-organizing system developments supporting real-time applications.

This book will aim to provide relevant theoretical frameworks and the latest empirical research findings in real-time self-organizing systems. The overall objective of this publication is to serve as a reference point for anyone engaging with real-time organic technologies from a theoretical, technical or application-oriented perspective.

Audience

This book is aimed at researchers in real-time embedded systems, particularly those who wish to understand the current state of the art in using Self-X capabilities in this domain. The book will serve primarily as a point of reference handbook to real-time organic-related technologies, applications and techniques, as well as an indicator of future and emerging trends to stimulate the interested readers. Researchers will also benefit from having such a reference book on their field, indicating the main achievements in the domain of distributed embedded and Self-X real-time systems as well its future trends and directions. Masters students will also find useful material that founds background reading for higher degrees in embedded and real-time systems.

Structure and Content

Using organic principles found in biology and nature is a key idea to handle the complexity of upcoming embedded systems, which will be distributed to many processors having multiple processor cores. Self-organization is a major concept in such organic principles. On the other hand, acting and reacting in a well-defined time (i.e., real-time behaviour) is an important requirement for many embedded systems. Bringing organic principles like self-organization and real-time behaviour together is a challenging task, which will be handled in the proposed book.

Real-time embedded systems are also themselves evolving. Distribution has always been an important consideration, but single processor nodes are rapidly being replaced by multithreaded microprocessor or multicore platforms. This has increased the emphasis on parallel programming and has had a major impact on embedded and real-time software development. The first chapter of this book, therefore, explores aspects of this issue. In Chap. 1, Brinkschulte et al. discuss how control theory can be used to improve the real-time capability of a modern multithreaded microprocessor by guaranteeing an instruction per cycle rate of a thread.

In general, executing critical services through small and portable devices in unbounded environments requires resilient and self-adaptive services and networks. These systems must be able to track the current status of the system, to estimate its uncertainties and to react accordingly. Due to the intrinsic asynchrony of the system and the complexity of its model, the achievement of the required dependability, timeliness and security is an important challenge. In Chap. 2, Bondavalli et al. called *Mobile Terminals* (MTs), developed as part of the ALARP[1] project.

[1] A railway automatic track warning system based on distributed personal mobile terminals.

Much of the work in adaptive distributed, embedded and real-time systems has focused on mobile services and *Ad Hoc Network* as the underlying base technology, and consequently many of the chapters in this book address issues with, or solve problems using, this framework. This issue has been particularly considered in Chaps. 3–6.

As a case study, a new methodology for real-time, self-organizing autonomous clustering for mobile ad hoc networks has been presented in Chap. 3. In this chapter, Kakuda, Ohta and Malek introduce a new concept of assurance networks and propose a unified design model and methodology, which is resilient to a diversity of node mobility and its variation. Within this model, each cluster head adjusts the interval of broadcasting a message within each cluster according to mobility of the cluster.

In Chap. 4, Khaluf, Mathews and Rammig present a novel approach for employing the swarm robotic techniques in real-time self-organized systems. This chapter extends existing time synchronization protocols for wireless communication and develops new strategies to use well-known time-synchronization protocols of wireless networks in swarm robotic communication. This work is based on two main strategies (i.e., *separations* and *connections*).

Controller Area Network (CAN) has been widely used in many distributed control applications. As embedded real-time systems grow in size, they will become more complex and have to cope with a mixture of hard and soft real-time systems. Scheduling these systems, within the imposed time constraints, is a challenging problem. The low capacity of CAN and usually strict constraints under which it is used in real-time applications limit its applicability. In Chap. 5,Ziermann, Salcic, and Teich consider a new approach for scheduling CAN messages, which is based on the online adaptation of message offsets, which results in significantly much higher utilization of CAN. The proposed approach is very simple and requires modest resources for its implementation, with no change in the standard CAN infrastructure.

In Chap. 6, Rettberg and Podolski presented a middleware architecture for distributed automotive systems that support self-configuration by dynamic load balancing of tasks. The inclusion of self-configurability is able to offer reliability within the multimedia network of the vehicle (Infotainment). Load balancing of tasks could be applied if an error occurred within the network. The error detection in the network and the load balancing should run automatically. Therefore, the middleware architecture has to deal, on the one hand, with the error detection and, on the other hand, with the migration of tasks. Additionally, to enable the migration it is important to identify the requirements of all electronic control units (ECU) and tasks within the network.

The successful development of autonomic systems requires new software development approaches. To increase the overall reliability, modern computerization of complex systems such as spacecraft, airplanes and automobiles strives to minimize the human control via self-adaptation. In such systems, special autonomic components are designed to handle performance problems via monitoring and automatic decision making. In Chap. 7, Higuera-Toledano reviews the state of the art in component-based real-time Java technology, considering both the problems and the solutions. In Chap. 8, Vassev and Hinchey present in detail a framework dedicated to the development of autonomic systems whereby developers are helped with problem formation, component and system design, analysis, evaluation and implementation. In the final chapter of this book, Brinkschulte et al. present aspects of emergence and an efficient implementation of an organic middleware based on an artifical hormone system to provide real-time task allocation in distributed and multicore systems.

Acknowledgements

The editors are grateful to Springer who gave them the opportunity to produce this book and to all the authors for agreeing to contribute in this work.

Madrid, Spain — M. Teresa Higuera-Toledano
Frankfurt am Main, Germany — Uwe Brinkschulte
Oldenburg, Germany — Achim Rettberg

Contents

1 **A Control Theory Approach to Improve Microprocessors for Real-Time Applications by Self-Adapting Thread Performance** ... 1
Uwe Brinkschulte, Daniel Lohn, Mathias Pacher, and Michael Bauer

2 **Providing Safety-Critical and Real-Time Services for Mobile Devices in Uncertain Environment** ... 25
Andrea Bondavalli, Francesco Brancati, Andrea Ceccarelli, and Lorenzo Falai

3 **Self-Organizing Real-Time Services in Mobile Ad Hoc Networks** ... 55
Yoshiaki Kakuda, Tomoyuki Ohta, and Miroslaw Malek

4 **Swarm Robotic Time Synchronization for Object Tracking** ... 75
Yara Khaluf, Emi Mathews, and Franz Josef Rammig

5 **Improving Performance of Controller Area Network (CAN) by Adaptive Message Scheduling** ... 93
Tobias Ziermann, Zoran Salcic, and Jürgen Teich

6 **Self-Configurable Automotive Embedded Systems** ... 119
Achim Rettberg and Ina Podolski

7 **Composing Adaptive Distributed Embedded and Real-Time Java Systems Based on RTSJ** ... 133
M. Teresa Higuera-Toledano

8 **The ASSL Formalism for Real-Time Autonomic Systems** ... 151
Emil Vassev and Mike Hinchey

9 **Organic Real-Time Middleware** ... 179
Uwe Brinkschulte, Mathias Pacher, Alexander von Renteln, and Benjamin Betting

Chapter 1
A Control Theory Approach to Improve Microprocessors for Real-Time Applications by Self-Adapting Thread Performance

Uwe Brinkschulte, Daniel Lohn, Mathias Pacher, and Michael Bauer

Abstract Modern superscalar, multithreaded microprocessors deliver high performance but lack of predictability. Our idea is to apply control theory on such a microprocessor to improve its real-time capability. The microprocessor is fitted with a closed control loop softening the impact of latency cycles and therefore stabilizing its throughput. Using a statistical processor model we are able to guarantee a defined rate of throughput and the stability of the controller. Additionally, we show the capability of the controlled microprocessor to save power and discuss the hardware overhead introduced by our closed control loop.

1.1 Introduction and Related Work

Nowadays, the development of embedded systems is strongly advancing. We find microprocessors embedded in all areas of life, e.g. in cell phones, cars and planes. In many of theses areas real-time capabilities are required to function faultlessly. However, modern microprocessors are designed to deliver a very high average performance rather than a high worst case performance. They use techniques, like caches or branch prediction, improving the average performance, but in the worst case they make it worse and hard to predict. Therefore, modern microprocessors are less suitable for embedded real-time systems, because it can be very hard guaranteeing execution time bounds.

The Worst Case Execution Time (WCET) analysis is a standard approach to determine, whether real-time systems can hold their constraints or not. The WCET

U. Brinkschulte (✉) • D. Lohn • M. Pacher • M. Bauer
Goethe Universität, Robert Mayer-StraSSe 11-15, Frankfurt am Main, Germany
e-mail: brinks@es.cs.uni-frankfurt.de; lohn@es.cs.uni-frankfurt.de; pacher@es.cs.uni-frankfurt.de; mbauer@es.cs.uni-frankfurt.de

M.T. Higuera-Toledano et al. (eds.), *Self-Organization in Embedded Real-Time Systems*,
DOI 10.1007/978-1-4614-1969-3_1,

analysis examines the semantic of the program code in respect of the processor used for its execution, resulting in a cycle accurate analysis.

An example of WCET analysis is the work in [16]. The authors discuss the WCET analysis of an out-of-order execution processor. They solve the WCET analysis problem by computing and examining the execution graph of the program code to be executed.

The authors of [6] consider WCET analysis of processors equipped with a branch predictor. They classify each control transfer instruction with respect to branch prediction and use a timing analyzer to estimate the WCET according to the instruction classification.

A good review on existing WCET analysis techniques is given in [14]. Additionally, a generic framework for WCET analysis is presented.

As we have seen, the WCET analysis can be very complex if the processor uses highly dynamic techniques, like out-of-order execution, jump prediction, etc.. Furthermore, the resulting time bounds may be far off the actual WCET.

So, we propose to use a self-adapting approach to simplify the WCET analysis. The microprocessor should be aware of occurring latency cycles and respond to them. We propose to use a control theory approach to improve the real-time capability of modern microprocessors and to provide better real-time guaranties. The microprocessor should be fitted with a closed control loop measuring and controlling the progress of each thread. The progress of a threads is given by its Instruction Per Cycle (IPC) rate. The benefit of this approach is an easier and tighter WCET analysis, because the controller automatically detects latency cycles and smoothes them out.

The throughput of a thread based on a statistical processor model is our transfer function in the sense of control theory. The statistical model has some similarities to processor performance prediction, but there are major differences. The main goal of processor performance prediction is to model the throughput of a processor as accurate as possible, but the predicted throughput may be higher as the actual throughput.

An example of performance prediction is the work of [13]. They use trace-derived information of different performance penalties and add up all those penalties to model the throughput of a superscalar processor. In [8] they propose a way of building up stacks to sum up the penalties for different miss events.

Whereas our statistical model also aims for an accurate prediction, the predicted throughput may not be higher as the actual throughput, since we are using it for real-time purposes. Furthermore, as shown in Sect. 6.4, inaccuracies of our model are reduced by the use of a controller. This is a main advantage of our approach. A first version of our statistical processor model was published in [5] and a more refined one in [17].

The idea to use closed control loops in real-time systems is not new. There are several approaches using it to limit communication overhead or to perform admission control, see [18] and [19]. These approaches are purely software-based. However, at the time we wrote this paper, to our knowledge no other research group is using a closed control loop to compensate latencies for real-time applications on a hardware level.

In IBM's Power 7 Chip, which supports simultaneous multithreading (SMT) on each core, enhanced SMT features are offered. They make a step towards controlling, by e.g. turning off threads if resources are getting rare or by using adaptive power management features, see [12].

Furthermore our approach is completely different to classical adaptive microprocessors, where resources like caches or queues are adapted at runtime, see [7, 10, 11] and [22].

So, our control theory approach contributes to the state of art in two ways: First, it shows how control theory can be used to improve the real-time capability of a modern multithreaded microprocessor by guaranteeing a certain IPC rate of a thread. Second, it shows how the improved execution time bounds can be used to save power.

This work is structured as follows: Sect. 6.1 gives an introduction and presents related work and similar approaches. In Sect. 6.2 we show that control theory can be used to improve the real-time capability of a multithreaded microprocessor in general. Section 6.3 presents the transfer function of the throughput of a thread based on a statistical processor model. And in Sect. 6.4 we give an upper and lower bound for the IPC rate while controlling, define the diameter of the convergence interval of the IPC rate. In Sect. 6.5 we evaluate our control theory approach by simulation. Section 6.6 considers an additional control variable, the processor frequency, in combination with the main control variable, the GP value to save power. In Sect. 6.7 we examine the hardware overhead introduced by our closed control loop. Section 1.8 concludes this work and gives an outlook to future work.

1.2 Basics

In this section we describe how control theory can be used to improve the real-time capability of a multithreaded microprocessor. The microprocessor is fitted with a closed control loop measuring and controlling the progress of each thread, see Fig. 1.1. The progress of a thread is given by its Instruction Per Cycle (IPC) rate. As scheduling strategy we choose Guaranteed Percentage Scheduling (GP scheduling, see [4]). In GP scheduling the processing time is partitioned into time periods (e.g. 100 clock cycles). A thread may now request a certain percentage of the time period (GP value, e.g. $0.6 \equiv 60\,\%$) and the corresponding number of clock cycles is guaranteed to that thread within the time period. The GP value of a thread can be changed at the end of each time period. Therefore, GP scheduling is able to control each thread in a fine-grained way and is used as control variable of the closed control loop.

If a microprocessor does not support GP scheduling, but offers dynamic priorities, the GP values in our approach can be mapped to those priorities. Each thread

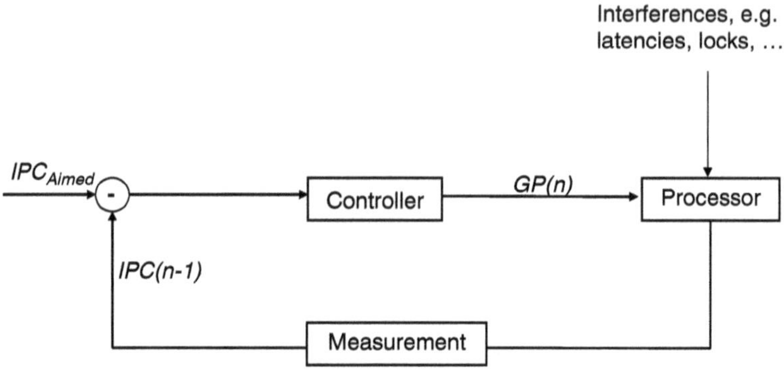

Fig. 1.1 A closed control loop

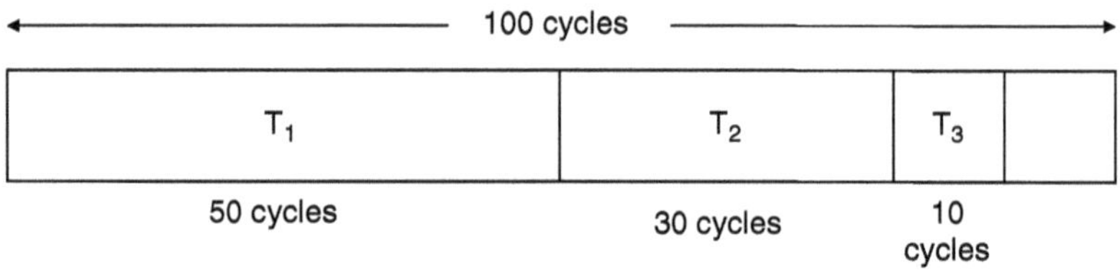

Fig. 1.2 Example of a GP schedule

receives, within a defined period of time, the highest priority for a time fragment equal to its GP value. Therefore, the resulting thread execution times are proportional to the GP values. Thus, our approach is not limited to GP scheduling, but can be adapted to dynamic priorities.

Example. Let T_1, T_2 and T_3 be three threads running on a processor with dynamic priorities. Let their GP values be 0.5, 0.3 and 0.1 and the time period 100 clock cycles. At first thread T_1 has the highest priority for $0.5 \cdot 100$ clock cycles = 50 clock cycles. Then thread T_2 has the highest priority for $0.3 \cdot 100$ clock cycles = 30 clock cycles. Finally, thread T_3 has the highest priority for $0.1 \cdot 100$ clock cycles = 10 clock cycles. After the entitled clock cycles have been granted to a thread, its priority is reduced to zero (Fig. 1.2).

The closed feedback controller measures the IPC rate of each thread at the end of a time period and compares it to the aimed IPC rate. Depending on the difference a new GP value is computed for each thread.

The throughput of a thread of a non controlled microprocessor in time period i is:

$$I(i) = T(i) \cdot G(i) \tag{1.1}$$

$G(i)$ is the assigned GP value and $0 \leq T(i) \leq 1$ can be seen as a throughput restraining factor caused by latencies like branch misprediction, cache misses, etc..

In case of a controlled microprocessor the closed feedback controller can compensate latency cycles of a thread. Considering a standard proportional controller (*P*) we get the following throughput of a thread of a controlled microprocessor in time period *i*:

$$I(i) = \frac{P \cdot T(i)}{1 + P \cdot T(i)} \cdot G(i) \tag{1.2}$$

As we can see in equation (1.2) the throughput converges to the aimed IPC value, that is given by $G(i)$, if *P* is chosen arbitrarily high. This shows that the impact of latency cycles represented in the throughput restraining factor $T(i)$ can be softened by the use of control theory.

1.3 Modeling

For applying control thory, a transfer function of the controlled entity is necessary to allow predictions. In this section we present such a transfer function of the throughput of a thread based on a statistical model of a multithreaded microprocessor. For this approach we consider a superscalar microprocessor with one core and *N* multithreaded pipelines. Furthermore, our microprocessor is equipped with caches and a branch prediction unit.

A microprocessor with *N* pipelines should reach, in theory, a throughput of *N* instruction per cycle, once the pipelines are filled. Unfortunately, the actual IPC rate is much lower due to throughput hazards. So we have to distinguish between useful clock cycles and latency cycles. Every instruction needs a certain amount of useful clock cycles to be executed, e.g. one clock cycle, plus an additional amount of latency cycles, if the instruction is afflicted with a throughput hazard.

Since we are using GP scheduling, we should calculate the throughput of a thread at the end of every time period. Therefore, we have to consider the number of instructions *m* of a thread to be executed within period *i*, the assigned GP value $G(i)$, the number of useful clock cycles C_{use} and additional latency cycles $C_{latency}$ in this period. Therefore, a thread running on a scalar microprocessor has a throughput of:

$$I(i) = \frac{m \cdot G(i)}{C_{use} + C_{latency}} = \frac{m \cdot G(i)}{m + C_{latency}} \tag{1.3}$$

Now lets consider a superscalar microprocessor with *N* pipelines, where the *m* instructions are equally distributed on all pipelines. Since each pipeline has to execute $\frac{m}{N}$ instructions, the throughput of such a single pipeline calculates to:

$$I(i) = \frac{\frac{m}{N} \cdot G(i)}{\frac{m}{N} + C_{latency_{pipe}}} = \frac{1}{N} \cdot \frac{m \cdot G(i)}{\frac{m}{N} + C_{latency_{pipe}}} = \frac{1}{N} \cdot \frac{G(i)}{\frac{1}{N} + \frac{C_{latency_{pipe}}}{m}} = \frac{1}{N} \cdot \frac{G(i)}{\frac{1}{N} + D(i)} \tag{1.4}$$

Combing the throughput of all N pipelines gives us the throughput of the thread running on a superscalar microprocessor:

$$I(i) = N \cdot \frac{1}{N} \cdot \frac{G(i)}{\frac{1}{N} + D(i)} = \frac{G(i)}{\frac{1}{N} + D(i)} \tag{1.5}$$

$D(i)$ is the average number of latency cycles per instruction and pipeline within interval i. It includes every occurring throughput hazard as well as additional latency cycles due to insufficient parallelism. $D(i)$ is a key element of the throughput restraining factor, see equation (1.1):

$$T(i) = \frac{1}{\frac{1}{N} + D(i)} \tag{1.6}$$

To simplify the calculation of $D(i)$ we divide hazard events into different classes. For each class j we model the hazard event by its frequency of occurrence $f_j(i)$ in interval i and its penalty value d_j. Equation (1.7) shows the calculation of the average rate of latency cycles per instruction $D_j(i)$ for a hazard class j based on these values. Instead of the frequency of occurrence $f_j(i)$, also the probability $p_j(i)$ of an instruction to be afflicted with a hazard of class j can be used.

$$D_j(i) = \frac{f_j(i)}{N \cdot m} \cdot d_j = \frac{p_j(i)}{N} \cdot d_j \tag{1.7}$$

Examples of those hazard classes are instruction and data cache misses, Branch Target Buffer (BTB) misses, mispredictions, thread switching penalties, dispatch inefficiency, as mentioned in [9], as well as insufficient parallelism.

To calculate the overall amount of latency cycles per instruction for all hazard classes j, we now simply have to add the individual values for each class, see equation (5.2).

$$D(i) = \sum_{j=1}^{n} D_j(i) \tag{1.8}$$

Applying equations (1.7) and (5.2) to equation (1.5) gives us the opportunity to check the plausibility this model:

$$I(i) = \frac{G(i)}{\frac{1}{N} + D(i)} = \frac{G(i)}{\frac{1}{N} + \sum_{j=1}^{n} D_j} = \frac{G(i)}{\frac{1}{N} + \sum_{j=1}^{n} \frac{f_j(i) \cdot d_j}{N \cdot m}} = \frac{G(i)}{\frac{1}{N} + \frac{1}{N \cdot m} \cdot \underbrace{\sum_{j=1}^{n} f_j(i) \cdot d_j}_{C_{latency_{all}}}}$$

$$= \frac{G(i)}{\frac{1}{N} + \frac{C_{latency_{all}}}{N \cdot m}} = \frac{G(i)}{\frac{1}{N} + \frac{C_{latency_{pipe}}}{m}} = \frac{G(i)}{\frac{1}{N} + D(i)} \tag{1.9}$$

Note: $\sum_{j=1}^{n} f_j(i) \cdot d_j$ is the sum of all occurring latencies in all pipelines.

This approach to model the throughput of a thread is not limited to a specific microprocessor. Its generic form, see equation (1.9), can be used for every microprocessor, if the amount of latency cycles are known or can be estimated.

1.4 Controlling

In this section we show the stability of a controlled microprocessor, give upper and lower bounds for the IPC rate and define the convergence interval of the IPC rate. First rather simple theoretical examinations have been published in [2, 3].

1.4.1 Defining the Transfer Function

The transfer function modeling a threads IPC rate is according to equation (1.9):

$$I(i) = \frac{g(i)}{\frac{1}{N} + D(i)} \tag{1.10}$$

The major difference to equation (1.9) is using $g(i)$ instead of $G(i)$. While $G(i)$ defines the guaranteed percentage value of a thread without any saturation effects, $g(i)$ takes these effects into account. Saturation occurs, because the sum of all GP values for all threads is limited to 1. Therefore we have the following in equation:

$$0 \leq g(i) \leq G(i) \leq 1 \tag{1.11}$$

The reaction of a proportional feedback controller can be smoothed by averaging the IPC rate of the last l periods instead of using the IPC rate of the last period only. Using this improvement, we get a slightly different transfer function:

$$I_l(i) = \frac{1}{l} \sum_{j=i-l+1}^{i} \frac{g(j)}{\frac{1}{N} + D(j)} \tag{1.12}$$

The difference Δ between two consecutive periods is given by equation (1.13). Δ consists of two different parts, firstly the gain of period i, and secondly the loss of period $i-l$.

$$\begin{aligned} I_l(i) &= I_l(i-1) + \Delta \\ \Delta &= \frac{1}{l} \frac{g(i)}{\frac{1}{N} + D(i)} - \Delta_2 \\ \Delta_2 &= \frac{1}{l} \frac{g(i-l)}{\frac{1}{N} + D(i-l)} \end{aligned} \tag{1.13}$$

We are able to give lower and upper bounds for Δ_2, because the guaranteed percentage value is always in the codomain of $[0\cdots 1]$, and the sum of all latency cycles per instruction $D(i-l)$ can not be negative:

$$0 \leq \Delta_2 \leq \frac{N}{l} \tag{1.14}$$

Using inequation (1.14) we are able to find the following bounds for Δ:

$$-\frac{N}{l} \leq \Delta \leq \frac{N}{l} \tag{1.15}$$

Therefore the variation of the average IPC rate after one period is limited by the number of simultaneous execution units N, and the chosen window size l.

Now we take a look at the controller and the creation of a new GP value. Considering a proportion controller with the factor P and an aimed IPC rate of I_A, we get equation (1.16). Taking saturation effects into account, we may have to subtract a certain amount $s(i)$, because the sum of the GP values of all threads is limited to 1.

$$g(i) = P(I_A - I_l(i-1)) - s(i) \tag{1.16}$$

Combining equations (1.16) and (1.13) results in the average IPC rate of the last l periods:

$$\begin{aligned}
I_l(i) &= I_l(i-1) + \frac{1}{l}\frac{P(I_A - I_l(i-1)) - s(i)}{\frac{1}{N} + D(i)} - \Delta_2 \\
&= I_l(i-1)\left(1 - \frac{P}{\frac{l}{N} + l\cdot D(i)}\right) + I_A\left(\frac{P}{\frac{l}{N} + l\cdot D(i)}\right) - \Delta_2 - \frac{s(i)}{\frac{1}{N} + D(i)} \\
&= (I_l(i) - \Delta)\left(1 - \frac{P}{\frac{l}{N} + l\cdot D(i)}\right) + I_A\left(\frac{P}{\frac{l}{N} + l\cdot D(i)}\right) - \Delta_2 - \frac{s(i)}{\frac{1}{N} + D(i)} \\
&= I_A - \Delta\left(\frac{\frac{l}{N} + l\cdot D(i)}{P} - 1\right) - \Delta_2\left(\frac{\frac{l}{N} + l\cdot D(i)}{P}\right) - \frac{s(i)}{\frac{1}{N} + D(i)}\cdot\frac{\frac{l}{N} + l\cdot D(i)}{P} \\
&= I_A - \Delta\left(\frac{\frac{l}{N} + l\cdot D(i)}{P} - 1\right) - \Delta_2\left(\frac{\frac{l}{N} + l\cdot D(i)}{P}\right) - s(i)\left(\frac{l}{P}\right)
\end{aligned} \tag{1.17}$$

1.4.2 Lower and Upper Bounds of the IPC Rate

Now we are able to consider lower and upper bounds for the average IPC rate of the last l intervals within interval i of the P-controlled processor. We start with the calculation of the lower bound and define it as $I_{l_{min}}$. We therefore combine equation (1.17) with inequations (1.15) and (1.14):

$$\begin{aligned} I_l(i) &\geq I_A - \frac{N}{l}\left|\frac{\frac{l}{N}+l\cdot D(i)}{P}-1\right| - \frac{N}{l}\left(\frac{\frac{l}{N}+l\cdot D(i)}{P}\right) - s(i)\left(\frac{l}{P}\right) \\ &\geq I_A - \left|\frac{1+N\cdot D(i)}{P}-\frac{N}{l}\right| - \frac{1+N\cdot D(i)}{P} - \frac{l}{P}\cdot s(i) \\ &=: I_{l_{min}} \end{aligned} \tag{1.18}$$

We find an upper bound defined as $I_{l_{max}}$ similar to inequation (1.18):

$$\begin{aligned} I_l(i) &\leq I_A + \frac{N}{l}\left|\frac{\frac{l}{N}+l\cdot D(i)}{P}-1\right| - s(i)\left(\frac{l}{P}\right) \\ &\leq I_A + \left|\frac{1+N\cdot D(i)}{P}-\frac{N}{l}\right| - \frac{l}{P}\cdot s(i) \\ &=: I_{l_{max}} \end{aligned} \tag{1.19}$$

1.4.3 *Defining the Convergence Interval of the IPC Rate*

Since we now have lower and upper bounds for the average IPC rate, we are able to define a convergence interval:

$$\begin{aligned} &I_{l_{max}}(i) - I_{l_{min}}(i) \\ &\qquad = I_A + \left|\frac{1+N\cdot D(i)}{P}-\frac{N}{l}\right| - \frac{l}{P}\cdot s(i) \\ &\qquad\quad -I_A - \left|\frac{1+N\cdot D(i)}{P}-\frac{N}{l}\right| - \frac{1+N\cdot D(i)}{P} - \frac{l}{P}\cdot s(i) \\ &\qquad = 2\cdot\left|\frac{1+N\cdot D(i)}{P}-\frac{N}{l}\right| - \frac{1+N\cdot D(i)}{P} \end{aligned} \tag{1.20}$$

It is interesting to see how the width of the convergence interval is affected by a growing value of P and $i \geq l$:

$$\lim_{P\to\infty, i\geq l}\left(I_{l_{max}}(i) - I_{l_{min}}(i)\right) = 2\cdot\frac{N}{l} \tag{1.21}$$

1.4.4 Stability and Saturation

As we can see in equation (1.21), the diameter of the convergence interval is independent of saturation effects, limited to a constant value, and as smaller as larger l, respectively P, is. Since we were able to find a constant limit for the convergence interval, even if P is chosen arbitrarily high, it also proves the stability of the controller. Therefore, a smaller convergence diameter can be achieved by a larger window size l and a larger value of P.

Now we take a look at the minimum IPC rate again. As we can see in equation (1.18), the number of latency cycles per instruction $D(i)$ is divided by the factor P of the controller. This means, inaccuracies caused by too pessimistic upper bounds for $D(i)$ are reduced by the use of a controlled processor. If, e.g. a value of 100 is chosen for P, the influence of all terms containing $D(i)$ is diminished by a factor of 100 as well. Therefore, the analysis and profiling to determine $D(i)$ for a controlled microprocessor can be simpler and more relaxed then for a non-controlled one.

If P is chosen arbitrarily high and saturation effects can be neglected, the lower bound of the IPC rate gets completely independent of $D(i)$ and can be determined by the simple result given in equation (1.22).

$$\begin{aligned} I_{l_{min}} &\geq \lim_{P \to \infty} I_A - \underbrace{\left| \frac{1+N \cdot D(i)}{P} - \frac{N}{l} \right|}_{\to \frac{N}{l}} - \underbrace{\frac{1+N \cdot D(i)}{P}}_{\to 0} - \underbrace{\frac{l}{P} \cdot s(i)}_{\to 0} \\ &= I_A - \frac{N}{l} \end{aligned} \tag{1.22}$$

In this case the difference between the aimed IPC rate and the guaranteed minimum IPC rate is constant and can be affected by the size of the control window l. Same is true for the upper bound, as can be easily derived from equation (1.19).

$$I_{l_{max}} = I_A + \frac{N}{l} \tag{1.23}$$

However, $D(i)$ influences the saturation effects. This can be shown by answering the following question: Under which conditions saturation effects will never occur? To calculate this, we can determine the maximum processor performance by using equation (1.10) and setting the sum of all $g(i)$ to 1. As a consequence, no saturation effects can occur if the sum of all aimed IPC rates of all threads holds the following inequation:

$$I_{saturation} \leq \frac{1}{n} \sum_{j=1}^{n} \frac{1}{\frac{1}{N} + D_{max}} = \frac{1}{\frac{1}{N} + D_{max}} \tag{1.24}$$

$D_{max} = max_{\forall i}(D_i)$ is the maximum value of the average latency cycles per instruction over all periods. As an example, if we assume 4 parallel pipelines ($N = 4$) and a maximum value of 2 for the average latency cycles per instruction ($D_{max} = 2$), saturation will definitely not occur as long as the requested IPC rate is below $I_{saturation} = 0.44$.

In fact, using D_{max} is very pessimistic, since the worst case value of $D(i)$ for all periods is assumed to be effective in each period. A better approach is to use the average latency value of all threads instead of the maximum amount. This allows a more realistic estimation for the price of calculating a proper value of D_{avg} by profiling or an analysis of throughput hazards.

$$I_{saturation} \leq \frac{1}{\frac{1}{N} + D_{max}} \leq \frac{1}{\frac{1}{N} + D_{avg}}$$

$$D_{avg} = \frac{1}{n} \sum_{j=1}^{n} D(j) \tag{1.25}$$

If saturation effects occur, $s(i)$ cannot be neglected. Fortunately like for $D(i)$, the influence of $s(i)$ is also diminished by the factor P, which can be seen in equations (1.18) and (1.19).

1.5 Evaluation

So far, we have theoretically calculated upper and lower bounds for the IPC rate of the controlled microprocessor, defined a convergence interval, shown stability and considered the effect of saturation. In this section we now simulate and evaluate our controlled microprocessor to validate the theoretical results. Therefore, we developed a statistical processor simulator. The simulator uses pseudo code instead of real code. This pseudo code only contains functional and statistical information, no real computation is done. The most realistic way to obtain pseudo code is to transform pipeline traces, e.g. derived from MSim [23]. MSim is a multithreaded Version of SimpleScalar. Another possibility to obtain pseudo code is to generate it by statistical data or on the basis of desired latencies.

The simulator is designed to support different threads and GP scheduling is used to isolate the threads from each other. The GP values for each thread can be either constant (the controller is not enabled), or self-adapted by a P-controller.

The simulator was validated by comparing its results to the results of the MSim simulator in the following way. First, a benchmark was run by MSim resulting in an IPC rate and a pipeline trace. The pipeline trace was analyzed and transformed into pseudo code for our simulator. Then, the pseudo code was run by our simulator, without controlling, also resulting in an IPC rate. The results were almost identical thus showing the usability of the simulator.

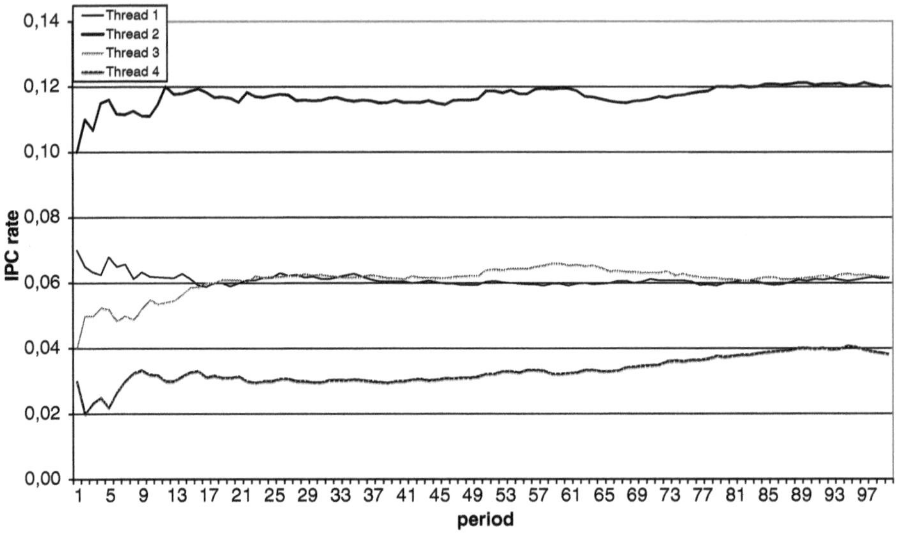

Fig. 1.3 Results without controlling

At first, we were running four threads with constant GP values, no controlling was done. The aimed IPC rate of the threads were:

- Thread 1: 0.1
- Thread 2: 0.2
- Thread 3: 0.1
- Thread 4: 0.05

As shown in Fig. 1.3, neither of them was able to reach its aimed IPC rate due to various throughput hazards. For each period, the displayed IPC rate is averaged on the preceding 50 periods to allow a comparison to the controlled processor. Each period consists of 100 clock cycles.

Thereafter, we were using the same four threads and aimed IPC rates as before, but enabled the controller. The controller used a control window size l of 50 periods and the proportional factor P was defined as 100. As we can see in Fig. 1.4, each thread is now able to reach its aimed IPC rate, once the control window is filled.

Now, we compared the results of the simulation with our theoretical model. In the simulation runs without controlling we found our throughput model being quite accurate, only a little bit too optimistic. The predicted IPC rate was about 3 % above the measured one.

Furthermore, our experiments showed an upper bound for the average number of latency cycles per instruction of $D_{avg} = 0.6$. Since the requested IPC rate for all four threads was 0.45 $(0.1 + 0.2 + 0.1 + 0.05)$, we can neglect saturation effects, because inequation (1.25) is true for those values.

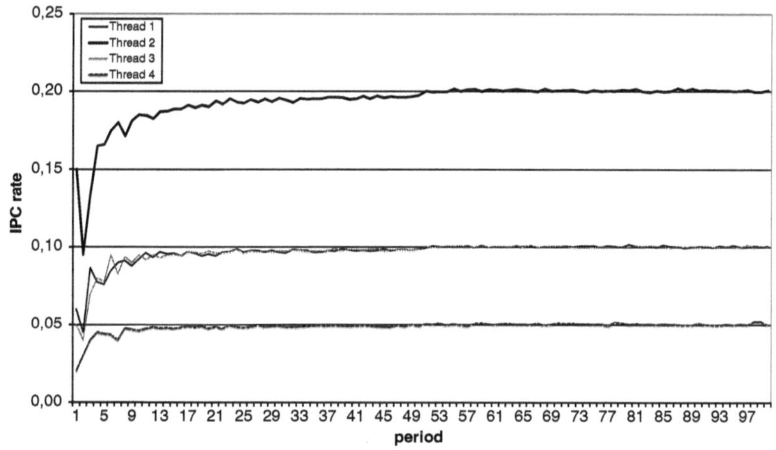

Fig. 1.4 Results of a controlled processor

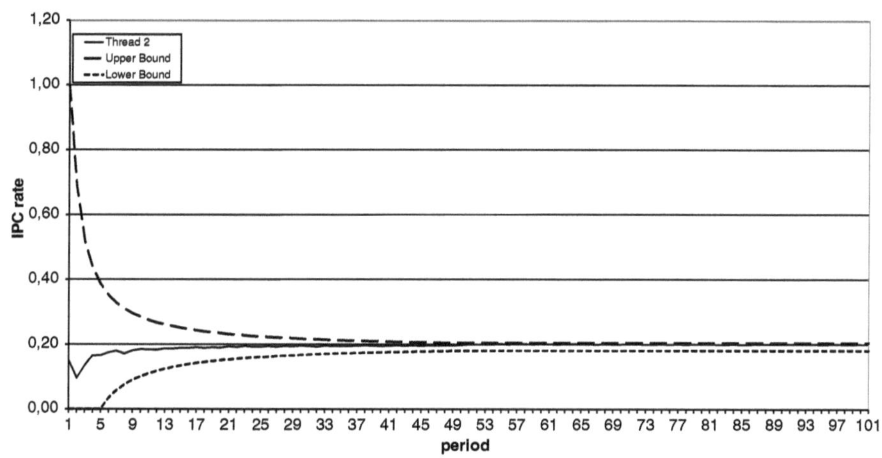

Fig. 1.5 IPC rate of Thread 2 with upper and lower bound

As an example, we calculated upper and lower bounds of the IPC rate of Thread 2 according to inequations (1.18) and (1.19). The results including the measured IPC rate are shown in Fig. 1.5.

We found an IPC rate of 0.180 as lower bound and an IPC rate of 0.204 as upper bound of Thread 2, once the control window was filled. The measured average IPC rate of Thread 2 was 0.197. Therefore the diameter of the convergence interval calculates to 0.024.

All other experiments made showed as well the validity of our theoretical results.

1.6 Frequency Scaling

Up to now we have showed that the predictability of thread execution can be improved by using closed control loops. Now we consider an additional control variable, the processor frequency, in combination with the main control variable used so far, the GP value. With this extension we can exploit processor idle times to decrease the processor frequency thus saving power.

1.6.1 Assumptions and Idea

We assume our multithreaded processor has to run k threads $T_1, \ldots, T_k$. Let the GP value of thread i be $g_i(n)$ in the period n ($0 \leq g_i(n) \leq 1$). In addition, we suppose $\sum_{i=1}^{k} g_i(n) \leq 1$ to be true in each period n as the processor cannot run the threads properly otherwise. These GP values refer to a reference frequency f_{ref} of the processor.

The basic idea is to control the processor frequency in combination with the GP value by the closed control loop.

Example 1. Let our processor run thread T with a reference frequency $f_{ref} = 10^6\,Hz$ and let T's GP value be $g = 0.3$. This means the processor has an idle time of 70 % per period which we can be used for frequency scaling: The processor executes $10^6 \cdot 0.3 = 300{,}000$ instructions/s. Now, let's assume we halve the frequency and run the processor with $f = 5 \cdot 10^5\,Hz$ and we double the GP value of T to 0.6. Obviously the processor executes the same $5 \cdot 10^5 \cdot 0.6 = 300{,}000$ instructions/s. This shows halving the frequency and doubling the GP value leads to the same throughput as before.

We can generalize the results of the example above. Let $f(n)$ be the processor frequency in the period n. Then, the new GP value $G_i(g_i(n), f(n), n)$ of thread i is computed as follows:

$$G_i(g_i(n), f(n), n) := g_i(n)\frac{f_{ref}}{f(n)} \tag{1.26}$$

As a result, $G_i(g_i(n), f, n) \cdot f(n) = g_i(n)\frac{f_{ref}}{f(n)} \cdot f(n) = g_i(n) f_{ref}$ is true which shows that a thread reaches the same throughput using the original frequency and GP value compared to the new frequency and the new GP value according to formula (1.26).

Thus, the main question is how to compute a minimal processor frequency allowing the threads to reach their aimed throughput by controlling. We propose two approaches in the following. Both of them are answering the question in a different way.

1.6.2 Meeting the Throughput of a Single Real-Time Thread

In this approach we assume w.l.o.g. that only thread T_1 must meet real-time constraints and therefore we only adjust its GP value $g_1(n)$ ($\neq 0$) and the GP values of $T_2, \ldots, T_k$ are not scaled according to the processor frequency. In order to guarantee the requested execution throughput of all threads the following condition must hold in each period:

$$G_1(g_1(n), f(n), n) + \sum_{i=2}^{k} g_i(n) \leq 1$$

Applying equation (1.26), this is equivalent to:

$$f(n) \geq \frac{g_1(n)}{1 - \sum_{i=2}^{k} g_i(n)} f_{ref}$$

Note that $1 - \sum_{i=2}^{k} g_i(n) > 0$ holds because $g_1(n) > 0$ holds. As a result we can set the minimal processor frequency $f_{\min}(n)$ in period n to:

$$f_{\min}(n) = \frac{g_1(n)}{1 - \sum_{i=2}^{k} g_i(n)} f_{ref} \tag{1.27}$$

Example 2. Let T_1, T_2, T_3 be three threads and let their GP values in the period n be $g_1(n) = 0.3$, $g_2(n) = 0.4$, and $g_3(n) = 0.1$ relative to the reference frequency $f_{ref} = 10^6\,Hz$. $g_1(n) + g_2(n) + g_3(n) = 0.8 \leq 1$ holds. According to equation (1.27) we compute the minimal frequency $f_{\min}$ to:

$$f_{\min}(n) = \frac{0.3}{1 - 0.4 - 0.1} \cdot 10^6\,Hz = 6 \cdot 10^5\,Hz$$

Example 3. Let again T_1, T_2, T_3 be three threads and let their GP values in the period n be $g_1(n) = 0.2$, $g_2(n) = 0.1$, and $g_3(n) = 0.1$ relative to the reference frequency $f_{ref} = 10^6\,Hz$. $g_1(n) + g_2(n) + g_3(n) = 0.4 \leq 1$ holds. Again according to equation (1.27) we compute the minimal frequency $f_{\min}$ to:

$$f_{\min}(n) = 2.5 \cdot 10^5\,Hz$$

Examples 2 and 3 show that we can choose the frequency as smaller as smaller the requested GP values (relative to f_{ref}) are. They are used to keep T_1's throughput constant although the frequency is decreased. This observation shows that frequency reduction is not possible if $\sum_{i=1}^{k} g_i(n) = 1$ holds because then the following equation holds according to equation (1.27):

$$f_{\min}(n) = \frac{g_1(n)}{1 - \sum_{i=2}^{k} g_i(n)} f_{ref} = f_{ref}$$

1.6.3 Meeting the Throughput of all Real-Time Threads

In this approach we assume the GP values of all threads $T_1, \ldots, T_k$ are adjusted to meet their real-time constraints. This means the throughput of all threads has to be constant. Therefore, the following inequation must hold in each period:

$$\sum_{i=1}^{k} G_i(g_i(n), f(n), n) \leq 1$$

Applying equation (1.26), this is equivalent to:

$$f(n) \geq \left(\sum_{i=1}^{k} g_i(n)\right) f_{ref}$$

Therefore, we can set the minimal processor frequency $f_{\min}(n)$ in period n to:

$$f_{\min}(n) = \left(\sum_{i=1}^{k} g_i(n)\right) f_{ref} \tag{1.28}$$

Example 4. Let T_1, T_2, T_3 be three threads and let their GP values in the period n be $g_1(n) = 0.3$, $g_2(n) = 0.4$, and $g_3(n) = 0.1$ relative to the reference frequency $f_{ref} = 10^6\,Hz$. $g_1(n) + g_2(n) + g_3(n) = 0.8 \leq 1$ holds. According to equation (1.28) we compute the minimal frequency $f_{\min}$ to:

$$f_{\min}(n) = 8 \cdot 10^5\,Hz$$

Example 5. Let again T_1, T_2, T_3 be three threads and let their GP values in the period n be $g_1(n) = 0.2$, $g_2(n) = 0.1$, and $g_3(n) = 0.1$ relative to the reference frequency $f_{ref} = 10^6\,Hz$. $g_1(n) + g_2(n) + g_3(n) = 0.4 \leq 1$ holds. Again according to equation (1.28) we compute the minimal frequency $f_{\min}$ to:

$$f_{\min}(n) = 4 \cdot 10^5\,Hz$$

Also in this approach it is easy to see that frequency scaling is not possible if the GP values are too high, i.e. if $\sum_{i=1}^{k} g_i(n) = 1$ holds because then the following equation is true according to equation (1.28):

$$f_{\min}(n) = f_{ref}$$

The examples show that the minimal frequency in this approach is greater than the minimal frequency in the last approach. The reason is that the idle times of the processor are split between all threads in the second approach while they can be used to satisfy the demand of only thread T_1 in the first approach. However, the advantage of the second approach is that the real-time constraints of all threads are met.

1.7 Implementation

In this section, we describe the implementation of an VHDL model of a microprocessor enhanced with a closed control loop. Furthermore, we examine the hardware overhead introduced by this closed control loop.

1.7.1 Basic Boundaries of the Implementation

To handle many threads and to ensure on-time execution for real-time applications, a fast high-end processor would be a simple solution. However, high-end processors consume too much energy for most embedded applications. Furthermore, due to techniques like caches and speculation their real-time performance is hard to predict.

So our solution is to optimize the performance and utilization of a smaller and simpler processor core by using an additional controller to implement the closed control loop. This controller reacts self-adaptive to unpredictable latencies caused by load and store instructions, branches or locks. By dynamically adjusting the rate of the real-time threads to the currently lowest necessary level, these real-time threads meet their deadlines while low priority non-real-time threads get as much processor time as possible.

On the other hand, the use of an additional controller increases chip size and energy consumption. The resources of the additional controller unit could also be used for a faster processor instead.

As the control approach works well in theory and simulation, see Sects. 6.4 and 6.5, we turn our attention to the energy and space consumption of the controlling unit in proportion to the total consumption of an unmodified core.

In order to obtain a significant result valid for a wide range of control algorithms, the implemented controller unit had to be complex and space-intensive. We decided to implement the **P**roportional/**I**ntegral/**D**ifferential (PID) algorithm as a superset of the P-controller addressed in Sect. 6.4. A PID-controller is more accurate, flexible, but as well more space-/energy-consuming as a pure P-controller. Therefore, all results of the PID-controller evaluation regarding additional space and energy consumption are all the more valid for less complex controlling algorithms like the proportional controller.

Furthermore, basic settings and requirements of the controlling unit have been evolved by a PID-controller implementation in a software processor simulator using different benchmarks, see [1, 20]. The resulting controller parameters, e.g. width of the controlling window or length of data types are used as settings for our controller implementation.

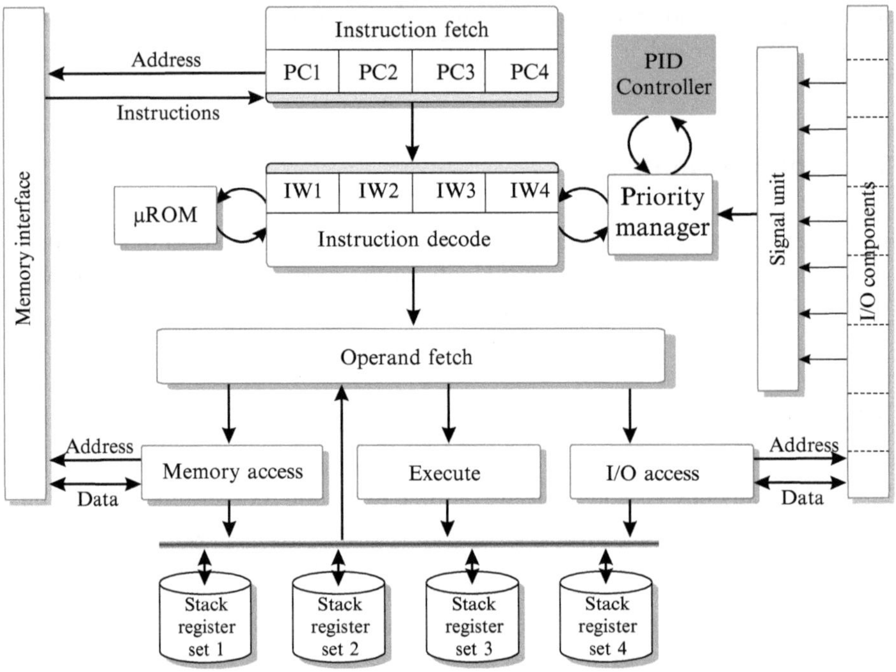

Fig. 1.6 The pipeline of the Komodo microcontroller

We used an VHDL implementation of a soft core to get a first impression of the increased core size. The additional used logic cells of an FPGA allows us a good estimation of increased energy and space consumption.

The soft core we used for this evaluation is an enhanced Komodo processor for embedded Java real-time systems. It offers the required capabilities for real time applications and is also multithreaded. Therefore, the mentioned goals of the additional controller can be tested on this processor.

1.7.2 The Komodo Microprocessor

The Komodo microprocessor is part of the Komodo microcontroller, which is described in [21]. For our purpose, only the microprocessor is important. Figure 1.6 shows the pipeline and the attached priority manager and signal unit. We enhanced it with an interface for our closed control loop. The interface effects the scheduling component only, leaving most of the core, especially the pipeline, untouched. The pipeline consists of the following four stages:

- Instruction fetch (IF)
- Instruction window and decode (ID)
- Operand fetch (OF)
- Execute and memory and I/O-access (EXE)

A detailed description of the pipeline stages and the processor core is given in [24].

The priority manager implements several real-time scheduling strategies, see [4, 15] and [24]. For our closed control loop only GP scheduling is required. The duration of a GP period as introduced in Sect. 6.2 for Komodo is 100 clock cycles. By monitoring the actual IPC rate, the GP value is adjusted by the newly introduced closed feedback loop. Occurring latency cycles reduces the IPC rate of a thread. Therefore, the controller has to increase the GP value of the thread to enable him to achieve its desired throughput.

In general, the actual IPC rate of a thread is defined as:

$$IPC = \frac{Number\ of\ instructions\ executed\ within\ time\ slice}{Duration\ of\ time\ slice\ (in\ clock\ cycles)} \tag{1.29}$$

Since Komodo is a scalar processor, each instruction takes one clock cycle to execute, except there is a latency. If we distinguish between active clock cycles, where an instruction of a thread is executed, and latency clock cycles, where nothing useful is done, we can refine the definition of the IPC rate for the Komodo microprocessor:

$$IPC = \frac{Number\ of\ active\ clock\ cycles\ within\ time\ slice}{Duration\ of\ time\ slice\ (in\ clock\ cycles)} \tag{1.30}$$

1.7.3 PID Controlling Algorithm

The general PID transfer function in its discrete form can be defined as follows:

$$y_n = K_P \cdot x_n + K_I \cdot \Delta t \cdot \sum_{v=1}^{n} x_v + \frac{K_D}{\Delta t} \cdot (x_n - x_{n-1}) \qquad \text{n} = 2,\ 3,4\ldots \tag{1.31}$$

Hereby, x_n is the difference at time n and y_n is the controller signal at time n. K_P, K_I and K_D are the controller constants while Δt is the duration between the measurement of x_n and x_{n-1}.

To get rid of the sum from 1 to n, there are two possible approaches. First, the recursive from can be calculated by subtracting y_n from y_{n-1}:

$$y_n = y_{n-1} + \left(K_P + K_I \cdot \Delta t + \frac{K_D}{\Delta t}\right) \cdot x_n + \left(-K_P - \frac{2 \cdot K_D}{\Delta t}\right) \cdot x_{n-1} + \frac{K_D}{\Delta t} \cdot x_{n-2} \tag{1.32}$$

Second, the non-recursive form can be used for the P- and D-part while only the sum is calculated recursively:

$$K_I \cdot \Delta t \cdot \sum_{v=1}^{n} x_v = K_I \cdot \Delta t \cdot \sum_{v=1}^{n-1} x_v + K_I \cdot \Delta t \cdot x_n \tag{1.33}$$

We have chosen the second possibility since it produces a bit less calculation effort and only the values x_n and x_{n-1} are needed.

The PID controller works in the processor as already shown in Fig. 1.1. The current IPC rate is computed by a measurement unit. The PID controller compares this value with the reference IPC value by computing their difference (x_n). Then the new GP value (y_n) of a thread is computed according to the PID algorithm given in equations (1.31) respectively (1.33).

Previous work on a PID controller for IPC rate stabilization was already done on a software processor simulator, based on the Kommodo architecture, see [20] and [1]. As a result, test values of the following essential parameters for the succeeding VHDL implementation have been determined:

- The values of K_P, K_I and K_D depend on the current running applications and the kind of occurring latencies. It was not part of the implementation work to find proper strategies to select this constants. So, values of prior simulation runs were taken for the VHDL implementation of the PID controller.
- The PID controller has to generate the new GP value for each thread every 100 clock cycles (the duration of a GP period).
- To prevent the controller from oscillation the IPC rate should be calculated for a longer time period, our controlling window. On the other hand, if the controlling window is chosen to big, the controller will react too lazy. Simulation results have shown that 1,000 clock cycles, equaling 10 controller or GP periods, are a very good value for the controlling window.

These boundaries define the structure and required space for the VHDL design.

Another advantage of the Komodo core was the reuse of the GP scheduler within the new control loop. The PID controller signal y_n directs the IPC rate via this scheduler. Therefore, the period the PID controller needs to generate a new signal is the same as the GP scheduling period. So the PID controller simply controls the allocated GP value within a range of 0 to 100 to reach the desired IPC rate. The former input of the GP scheduler is used as set value for the PID controller and the signal y_n forms the new input of the scheduler.

1.7.4 Resulting Processor Size

The evaluation of memory space needed for history tables and calculation units depends on the characteristics of the transfer function of the control algorithm. The GP scheduling period of 100 clock cycles also dictates the frequency of the PID controller. In order to determine the minimal required extra hardware, the upcoming operations has to be analyzed and distributed to additional calculation units. As these units have always 100 clock cycles to finish the calculation, the operations can be spread over the whole scheduling windows to minimize the calculation hardware needed at one time.

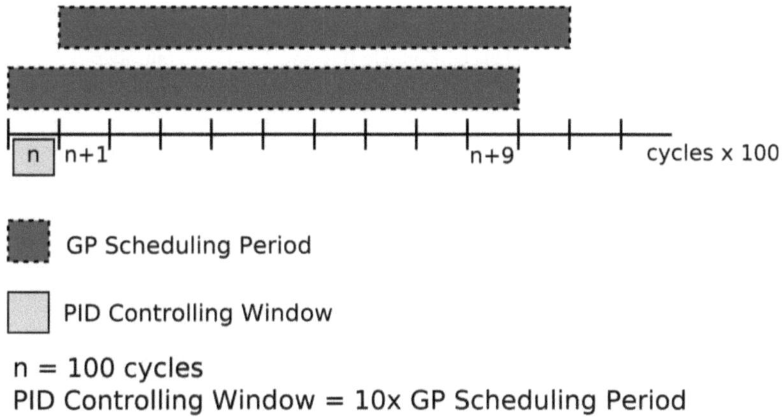

Fig. 1.7 Relation of PID controlling window and GP scheduling period

The analysis of the upcoming operations in one PID controller period of 100 clock cycles is:

- Calculating the average difference value x_n for the last 1,000 clock cycles. This period corresponds to 10 GP periods. So, a history containing the difference between actual and set values of the last 9 GP periods is stored and the arithmetical average of these 10 values is built, see Fig. 1.7.
- To calculate the P component of the PID controller transfer function, $K_P \cdot x_n$, one multiplication has to be done.
- To calculate the D component of the PID controller transfer function, $\frac{K_D}{\Delta t} \cdot (x_n - x_{n-1})$, one multiplication of the constant value $\frac{K_D}{\Delta t}$ by the result of $x_n - x_{n-1}$ has to be done. Regarding the way x_n was calculated, the difference of $x_n - x_{n-1}$ can be calculated by simply subtracting the difference between the 10th oldest GP scheduling period and the current GP scheduling period. So, the history mentioned above has to be expanded to the last 10 GP scheduling periods and only one subtraction of the current and the oldest history entry has to be done to get the result of $x_n - x_{n-1}$.
- To calculate the I component of the PID controller transfer function, the recursive approach shown in (1.33) is used. The current value of x_n has to be multiplied with the constant value $K_I \cdot \Delta t$. The result has to be added to the previous sum.

That enumeration showed the price of the PID controller we have to pay. The the number of calculations is quite low while the space needed to save the computation history is moderate.

An implementation of the PID controller on an Altera Cyclon II FPGA board showed the following result, see Table 1.1: The required "dedicated logic register" increased by 15 % while the number of "combinational functions" were nearly constant. The number of total logic cells increased by 6 % overall.

Table 1.1 Evaluation of the PID controller hardware overhead

Enhanced Komodo IP Core:	Without PID controller	With PID controller
Dedicated logic registers	4,688	5,412
Total combinational functions	12,696	12,830
Total logic elements	12,895	13,741
Space used on the Cyclon II FPGA	39 %	41 %

This is a satisfying result for a complex control unit. Considering the structure of more simple controlling algorithms, e.g. a proportional controller, having less calculation cost and memory requirements, these results were very promising.

The approach to minimize the PID calculation units can also be extended to use the existing hardware to control several threads at the same time. If the scheduling window is long enough the calculation of two or more threads can be executed pipelined without further additional hardware.

1.8 Conclusion and Future Work

We showed that control theory can be used to soften the impact of latency cycles to the throughput of a microprocessor, thus improving its real-time capability, see equation (1.2). Then, we presented a statistical model of a superscalar, multithreaded microprocessor to calculate the throughput of a thread, see equation (1.9). We considered different classes of throughput hazards, like cache misses, misprediction, etc. This model is a generic one and can be adapted to different processor architectures.

Furthermore, we were able to guarantee a lower and upper bound for the IPC rate while controlling, see inequations (1.18) and (1.19). Additionally, we defined the diameter of the convergence interval of the IPC rate, see equations (1.20) and (1.21). The stability of the controller was shown in equation (1.21).

To neglect saturation, the desired throughput of the microprocessor must hold inequation (1.25).

Since the number of latency cycles per instruction $D(i)$ in any period i is divided by the factor P, see equation (1.18), the effect of inaccuracies of the used processor model are reduced. Therefore, such an analysis is simplified and improved.

We showed, that controlling not only the GP value, but as well the processor frequency is also possible thus saving power. Here, the closed control loop exploits idle cycles to reduce the processor frequency.

Then, we examined an VHDL implementation of a soft core and considered additional hardware overhead introduced by our control loop. As we saw in Table 1.1 the overhead can not be neglected, but is quite low.

Future work will concern further improvement of the statistical model and the control theory approach, taking into account more advanced hardware techniques, like multicore processing and consider other programming models.

This research is done in frame of the REMIS project granted by the German Research Foundation (DFG) on 2009.

References

1. Brinkschulte U, Pacher M (2005) Improving the real-time behaviour of a multithreaded Java microcontroller by control theory and model based latency prediction. In: WORDS 2005, Tenth IEEE international workshop on object-oriented real-time dependable systems, Sedona, Arizona
2. Brinkschulte U, Pacher M (2008) A control theory approach to improve the real-time capability of multi-threaded microprocessors. In: ISORC, 2008– 11th IEEE International Symposium on Object/component/service-oriented Real-time distributed Computing, Orlando, Florida, USA, May 5–7, 2008, 399–404
3. Brinkschulte U, Pacher M (2009) A theoretical examination of a self-adaptation approach to improve the real-time capabilities in multi-threaded microprocessors. In: SASO 2009 – Third IEEE international conference on self-adaptive and self-organizing systems, San Francisco, California
4. Brinkschulte U, Ungerer T (2007) Mikrocontroller und mikroprozessoren, 2nd edn. Springer, Berlin
5. Brinkschulte U, Lohn D, Pacher M (2009) Towards a statistical model of a microprocessor's throughput by analyzing pipeline stalls. SEUS 2009 – 7th IFIP workshop on software technologies for future embedded and ubiquitous systems (SEUS) Newport Beach, USA, Nov 16–18, 2009
6. Colin A, Puaut I (2000) Worst case execution time analysis for a processor with branch prediction. Kluwer, Norwell, pp 249–274. http://dx.doi.org/10.1023/A:1008149332687
7. Corporation I (2002) Intel speedstep^{TM} technology. Intel Cooperation, Technical report
8. Eyerman S, Eeckhout L, Karkhanis T, Smith JE (2006) A performance counter architecture for computing accurate cpi components. SIGOPS Oper Syst Rev 40(5):175–184. http://doi.acm.org/10.1145/1168917.1168880
9. Eyerman S, Eeckhout L, Karkhanis T, Smith JE (2009) A mechanistic performance model for superscalar out-of-order processors. ACM Trans Comput Syst 27(2):1–37. http://doi.acm.org/10.1145/1534909.1534910
10. Gonzalez R, Horowitz M (1996) Energy dissipation in general purpose microprocessors. IEEE J Solid-State Circ 31:1277–1284
11. Hughes C, Srinivasan J, Adve S (2001) Saving energy with architectural and frequency adaptations for multimedia applications. In: Proceedings of the 34th Annual International Symposium on Microarchitecture. ACM/IEEE, Austin, Texas, USA, Dec 1–5, 2001
12. Kalla R, Sinharoy B, Starke WJ, Floyd M (2010) Power7: Ibm's next-generation server processor. IEEE Micro 30(2):7–15. http://dx.doi.org/10.1109/MM.2010.38
13. Karkhanis TS, Smith JE (2004) A first-order superscalar processor model. SIGARCH Comput Arch News 32(2):338. http://doi.acm.org/http://doi.acm.org/10.1145/1028176.1006729
14. Kirner R, Puschner P (2005) Classification of WCET analysis techniques. In: Proceedings of the 8th IEEE international symposium on object-oriented real-time distributed computing, pp 190–199
15. Kreuzinger J (2001) Echtzeitfähige ereignisbehandlung mit hilfe eines mehrfädigen java-mikrocontrollers. Ph.D. thesis, Logos Verlag, Berlin
16. Li X, Roychoudhury A, Mitra T (2004) Modeling out-of-order processors for software timing analysis. In: RTSS '04: Proceedings of the 25th IEEE international real-time systems symposium. IEEE Computer Society, Washington, DC, pp 92–103. http://dx.doi.org/10.1109/REAL.2004.33

17. Lohn D, Pacher M, Brinkschulte U (2011) A generalized model to control the throughput in a processor for real-time applications. In: 14th IEEE international symposium on object/component/service-oriented. Real-time distributed computing–ISORC 2011, Newport Beach, USA, Mar 28–31, 2011
18. Lu C, Stankovic JA, Tao G, Son SH (1999) Design and evaluation of a feedback control EDF scheduling algorithm. In: Real time systems symposium (RTSS) 1999, Phoenix
19. Lu C, Stankovic JA, Abdelzaher TF, Tao G, Son SH, Marley M (2000) Performance specifications and metrics for adaptive real-time systems. In: Real time systems symposium (RTSS) 2000, Orlando
20. Pacher M, Brinkschulte U (2005) Implementing control algorithms within a multithreaded java microcontroller. In: ARCS 05, 18th international conference on architecture of computing systems, Innsbruck
21. Pfeffer M, Ungerer T, Uhrig S, Brinkschulte U (2002) Connecting peripherial interfaces to a multi-threaded java microcontroller. In: Workshop on java in embedded systems, ARCS 2002
22. Sasanka R, Hughes C, Adve S (2002) Joint local and global hardware adapations for energy. In: Proceedings of the ASPLOS-X. ACM, New York
23. Sharkey JJ, Ponomarev DK (2005) M-sim: A flexible, multithreaded architectural simulation environment. Techenical report, Department of Computer Science, State University of New York at Binghamton
24. Uhrig S, Liemke C, Pfeffer M, Becker J, Brinkschulte U, Ungerer T (2002) Implementing real-time scheduling within a multithreaded java microcontroller. In: 6th workshop on multithreaded execution, architecture, and compilation MTEAC-6, Proceedings of the 35th Annual International Symposium on Microarchitecture. ACM/IEEE, Istanbul, Turkey, Nov 18–22, 2002

Chapter 2
Providing Safety-Critical and Real-Time Services for Mobile Devices in Uncertain Environment

Andrea Bondavalli, Francesco Brancati, Andrea Ceccarelli, and Lorenzo Falai

Abstract Executing critical services through small, portable devices in harsh environment with uncontrolled system boundaries requires resilient and self-adaptive services and networks, which are able to track the current status of the system, to estimate its uncertainties and to react accordingly. Achieving the required dependability, timeliness and security levels is an architectural challenge due to the intrinsic asynchrony of the system and the complexity of its fault model. In this paper we present the design of a resilient middleware and architecture for real-time, safety-critical and distributed mobile devices which operates in an uncertain environment. Such mobile devices, called Mobile Terminals (MTs), are developed as part of the ALARP (A railway automatic track warning system based on distributed personal mobile terminals) project which aims to design and develop a safety-critical Automatic Track Warning System (ATWS) for railway trackside workers. The MTs are wearable devices which operate in a railway worksite, to execute the real-time and safety critical service of notifying to the railway worker a train approaching the worksite or the occurrence of hazardous situation (e.g., workers missing or not responding).

2.1 Introduction

Trends on future networking systems include convergence towards pervasiveness and heterogeneity, where a multitude of (small, portable) nodes and sensors interacts among themselves and with infrastructures to execute services [36].

A. Bondavalli (✉) • A. Ceccarelli
Department of Systems and Informatics, University of Firenze, Viale Morgagni 65, Italy
e-mail: bondavalli@unifi.it; andrea.ceccarelli@unifi.it

F. Brancati • L. Falai
Resiltech S.r.l., Piazza Nilde Iotti 25, Pontedera (Pisa), Italy
e-mail: francesco.brancati@resiltech.com; lorenzo.falai@resiltech.com

M.T. Higuera-Toledano et al. (eds.), *Self-Organization in Embedded Real-Time Systems*,
DOI 10.1007/978-1-4614-1969-3_2,

The dynamic of the system, the infrastructure and the environment require attentive monitoring and self-adaptive solutions, as often system boundaries are unknown, or uncertainties affect the system and the activities performed on it (for example, threat and fault definition and occurrence rate may undergo a permanent update). This calls for the need of solutions to monitor the surrounding conditions and adapt accordingly.

As an example of the variability and uncertainties which characterize these systems, we can mention real-time and adaptive applications in *Wireless Sensor Networks* (WSNs) [35], for example sensing applications: these applications require a specific synchronization uncertainty in order to collect reliable data, but keeping clocks continuously synchronized brings relevant energy consumption. Thus, adaptive requirements for synchronization uncertainty may be established, with the aim to achieve an optimal tradeoff between the energy costs of transmitting synchronization beacons and the need to contain synchronization uncertainty within system requirements to collect reliable data [35]. We can also relate this tradeoff between energy costs and the service provided to the problem of accurate positioning and self-localization, which is nowadays required by many different kind of systems. Accurate localization in open space for small, portable devices is a hard task; for example the *Global Positioning System* (GPS) [18] signaling system *(i)* usually guarantees only positioning with meter-accuracy, *(ii)* is energy consuming, and *(iii)* may be unavailable (e.g., under thick foliages or in canyon). This implies that the quality of the position is subject to many different conditions, and services which relies on position estimation needs to be aware and to adapt to this variability.

In this paper we are targeting the execution of safety critical and real-time services through small, portable devices in harsh environment with uncontrolled system boundaries, which requires solutions for resilience and self-adaptiveness of services to the changing environment and operating conditions. We present a resilient solution for mobile devices, the *Mobile Terminals* (MTs), which are wearable, mobile devices that operate in a distributed, real-time system with strict requirements on safety and reliability. The MTs are developed as part of the ALARP project [1] whose goal is the study, design and development of a safety-critical (its Safety Integrity Level shall be at least equal to 2 according to the railway standards [17]) *Automatic Track Warning System* (ATWS) for railway trackside workers. Primary goal of an ATWS is to recall the attention of a workgroup operating on a railway worksite about the presence of a train approaching the worksite.

Several constraints arise when operating in a worksite-like environment, as unpredictable variation in network quality, lowered trust and robustness of mobile elements, limitations on local resources imposed by weight and size constraints, concern for battery power consumption, and uncertainties in time and position estimation [36]. Finally, concerns are on their non functional requirements, mainly when relating to dependability, security and real-time requirements. Achieving the required dependability, trustability and security levels despite accidental and malicious faults in networks composed by these mobile nodes is an architectural challenge.

In this paper we focus on the resilient solutions that are applied in the MTs architecture. We present the hybrid architecture system model adopted to distinguish the architecture of the single MT in two parts, characterized by different requirements in terms of timeliness and security, and then we describe the MT resilient middleware, which manages the underlying hybrid architecture and provides the dependability services. Attention is devoted to the services of the MT which aim to estimate the uncertainties in the system (time and localization uncertainty, and low-power criticality), and allow the system to react accordingly. Finally, a brief discussion on the envisioned solutions for the prototype is presented.

The remaining part of this paper is organized as follows. In Sect. 2.2 we present an overview of the ALARP project and of the key requirements identified. In Sect. 2.3 we present a high level description of the MT, underlying the need for self-adaptive and resilient services, and consequently of a resilient middleware. Sections 2.4–2.6 focus on the description of the MT middleware: the finite state machine of the middleware, the wormhole and the payload part of the MT and related middleware services are presented. Then, in Sect. 2.7 indications on the envisioned MT prototype, that shall be built at the end of the ALARP project, are provided. Finally, conclusions are in Sect. 5.6.

2.2 Overview and Key Requirements of ALARP

In this Section we present an overview of the ALARP system and its main requirements; a particular attention is devoted to the analysis of the dependability and real-time requirements.

2.2.1 Overview of ALARP

Safety of workers in the railway scenario is a serious concern; in fact, trackside workers are exposed to injuries and fatalities since vehicles are constrained to tracks and therefore drivers have much less margins to react in case of emergencies. For example, as reported in [19] for U.S. railways, during 1993–2002 there were 460 fatal railroad-related work injuries within railroading workers and another 761 fatal railroad-related work injuries involving workers which are not from the railroad field (e.g., those happening to workers in rail transportation occupations outside railroading or to truck-drivers in other industries who perish in at-grade crossing collisions with trains [19]).

The ALARP project [1] proposes to design, develop and validate an ATWS to improve the safety of trackside workers. The ALARP ATWS targets two key objectives. The first one is being able to inform the trackside workers about approaching trains on the track, emergencies on tracks and tunnels nearby the workers (e.g., fires in a tunnel, or toxic smoke), escape routes in case of emergencies, maintenance events on power lines and/or safety equipment in the concerned tracks that may put at risk workers' safety (e.g., being hit by a train or by an electric shock). The second is to monitor the status of the worker (through health monitoring) and his position, to identify the workers at risk (e.g., not responding due to health problems, or located close to the track while a train is approaching). Main requirements of the ALARP architecture and its components are:

- Self-powered devices not requiring external power supply;
- Portable and wearable devices communicating by using a wireless link;
- User friendly interfaces for low specialization workers;
- Flexibility in order to cover different working environments such as open line, stations, tunnels, bridges;
- Adaptability to all different working conditions (e.g., night or daytime) and weather conditions (e.g., fog, snow, rain, high/low temperature);
- Low cost and as much as possible based on Commercial Off-the-Shelf (COTS) components to be competitive in costs and sustainable for low and medium scale production;
- Capability of operating without interfering at all with existing signaling systems as for example the *Global System for Mobile communication-Railway* (GSM-R);
- Dependability and real-time requirements which constitute a critical part of the requirements set, as the notification of critical events shall be safely and timely delivered to the various MTs.

Before introducing the ALARP system, few notions are required on the definitions of railway worksites.

A worksite is an area on the railway on or close to a track where workers are acting; its typical maximum length (i.e., the dimension along the railway line) is 700 m. The worksite is divided in *red zones* and *green zones*. A red zone is not protected from rolling stock movements (i.e., it is typically an area that contains the track itself, where trains may pass).

Contrary to the red zone, a green zone is defined as a worksite on the track in which there are no rolling stock movements. It constitutes a safe area where workers move when a train is passing through the worksite.

The safe working limit in the red zone varies according to the different national regulations (and depends on the train speed and on line characteristics). For example, the safe working limit is 2.0 m (measured from the inner side of the rail) in UK [33], 1.9 m (but measured from the centre of the track) in Germany [24], and 1.5 m (measured from the inner side of the rail) in Belgium [20].

The ALARP system notifies working gangs along train lines providing *alerts* or *warnings* related to approaching trains and/or on-track plants in order to improve safety by supporting (or replacing) human lookouts. When workers' lives are at risk, an alert signal is sent to give them at least enough time to reach a safe position, according to the national regulations, otherwise warning signals are sent only to raise their attention [2, 4].

The alert time for workers in red zone is, according to the different national regulations, either the minimum time needed by the workers' gang to reach a

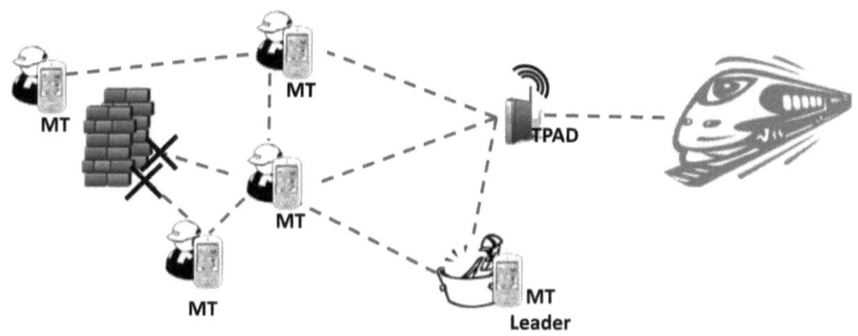

Fig. 2.1 The ALARP context and environment

position of safety or the minimum time interval between the alert signal and the train entering the worksite which gives everyone in the group sufficient time to reach the position of safety i.e., to reach a green zone. Note that these two definitions are coincident in terms of alert time. The duration of such alert time varies from country to country according to national regulations, for example it is 10 s in UK [33], 15 s in Belgium [20] and 35 s in Germany [24].

All elements are now set to present an overview of the components of the ALARP system. The key components are the *Train Presence Alert Devices* (TPADs) and the MTs, as shown in Fig. 2.1 and described below.

The trackside TPADs are able to sense approaching trains on the interested track, using long-range multi-spectral cameras and eaves-dropping the railway network communication. The TPADs are able to detect approaching trains in all lighting and weather conditions, they are self-powered, portable and easily movable in different places of work, and equipped with an anti-theft system. The TPADs are designed to be autonomous, self-powered alerting systems that serve a virtual gate and deliver an alarm to all the people within the relevant site (more precisely, to the MTs that the people in the working site are wearing) through a dedicated wireless channel in order to provide sufficient time to reach the green zone [37].

The MTs are instead distributed, low cost, wearable, context-aware, robust, trustable and highly reliable wireless devices, that inform the workers about possible approaching trains and/or other events that could put at risk their safety. The Mobile Terminal MT is able to generate alarms, and to communicate and interact through wireless connections with other MTs and the trackside TPADs together with mechanisms to check validity and trust levels for this ad-hoc communication.

2.2.2 *Resilience and Real-Time Requirements in the MT*

From the description of the ALARP system that we provided above, it is easy to note that the MT must guarantee requirements as resilience, security and trustability, despite the possible harsh conditions (e.g., obstacles, high noise and interferences, low light, etc.). Also, it is part of a distributed system in which nodes exchange real-time and safety-critical messages.

Achieving the required dependability and security levels despite accidental and malicious faults in networks composed by mobile nodes is particularly architectural challenging due to their intrinsic asynchrony (unreliable communication, partitioning, mobility, etc.). Several factors induce asynchrony in mobile systems based on wireless communication between them: the unreliability of the communication, the network partitioning, the changes in the network topology and the consequent absence of continuous connectivity to global resources [45]. Another important consideration is that coordinating the set of MTs and achieving hard real-time requirements in wireless communication may require that the MTs share a unique global time (we remark that GPS time-signaling techniques may not always be available e.g., in tunnels, or under thick foliage), thus requiring additional synchronization and localization mechanisms and means to take into account of possible uncertainties and to adapt to the different operating conditions. Precise positioning of the MTs is a challenging task, and the uncertainties introduced in the position computation, variable depending on the operating conditions and the surrounding environment, need to be estimated carefully and the solutions for position signaling and reaction (identify if the worker is in red or green zone) adapted accordingly [11].

Finally, we recognize that the threats to resilience and safety are particularly severe. Device lifetime and communication are severely limited by scarcity of power; the usage of wireless links means susceptibility to link attacks ranging from passive eavesdropping to active impersonation, message replay, and message distortion; poor physical protection of mobile devices (especially in a hostile environment like the one considered in ALARP) makes them susceptible to physical damage, and vulnerable to theft.

2.2.3 Safety Standards and MT Requirements

In this section we briefly describe the main requirements presented in the CENELEC standards that need to be satisfied to develop a SIL 2 compliant component [13–17]. The MT middleware and architecture has been conceived in order to meet these requirements.

Regarding *hardware requirements*, the standards require assurance that no single random hardware component failure mode is hazardous. This fail safety property can be achieved by *(i)* composite, *(ii)* reactive or *(iii)* inherent fail-safety techniques.

Composite fail-safety means that each safety-related function is performed by at least two independent items and non-restrictive activities are allowed only if the necessary number of items agree. Since the MT should be a component of reduced dimension, also the number of hardware components should be reduced to a minimum. That is why the possibility of composite fail-safety and of using hardware redundancy is limited.

Reactive fail-safety assures safe operation by proper detection and negation of hazardous faults that occur in a single item. However, the fault detection function is regarded as a second item that shall be independent in order to avoid common-cause failures. Independence could be lost by physical internal/external influences, or functional internal/external influences.

Inherent fail-safety is considered if all non-negligible failure modes of a single item are non-hazardous. Inherent fail-safety can be used for certain functions (e.g. to enforce shutdown). Justification shall be provided for any failure mode which is considered to be incredible.

The following techniques are highly recommended:

- Dynamic fault detection: online dynamic testing should be performed to check the proper operation of the safety-related system and provide an indication to the operator;
- Program sequence monitoring: temporal or logical monitoring of the program sequence and indication of faults to the operator;
- Measures against voltage breakdown, voltage variations, overvoltage, low voltage and temperature increase (to detect over temperature).

The standard mentions, in the case of reactive fail-safety, the following fault detection techniques: *(i)* encoding, *(ii)* multiple computation and comparison, and *(iii)* continual testing. In the case of the MT, the reactive fail-safety principle was applied, using a single electronic structure with self-test and supervision. Also, a large set of state-of-the art software fault detection techniques, suggested by the standards, are applied.

We observe that the particular conditions in which the MT is expected to operate required to slightly deviate from a traditional, state of the art approach on reactive fail safety using single electronic structure. That is, in order to improve guarantees on time and security requirements, still avoiding a fully redundant hardware architecture (to not impact on costs and size of the device), an approach based on the hybrid system model is applied, where different parts of the system are able to provide different guarantees on the critical requirements, and it shall be explained in the following of this paper.

2.3 The MT and Its Need of a Resilient Middleware

The MT is a safe component (at least SIL 2 according to CENELEC railway standards) and, to keep costs low, it shall be based on COTS hardware components.

The characteristics of the ALARP system and environment make necessary to adopt in the MT solutions for high resilience, including mechanisms to adapt to and mitigate the uncertainties of the system (mainly in terms of temporal and localization uncertainty). A *management middleware* (or simply *middleware*) was deemed necessary to guarantee the required safety critical and resilient properties. In the design phase, to separate different parts of the system that are characterized

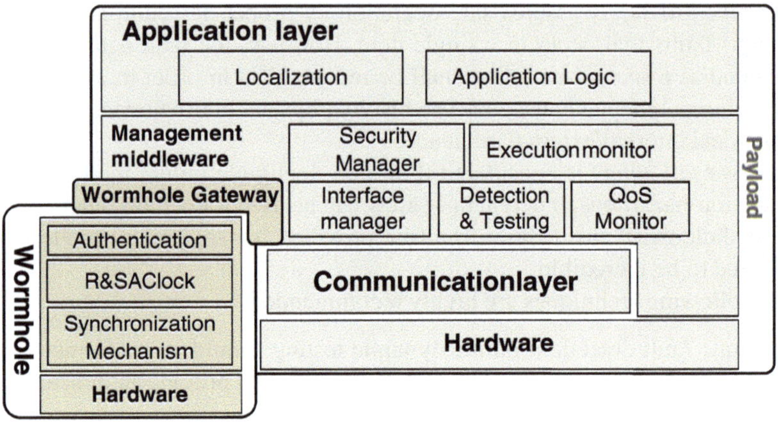

Fig. 2.2 Overall view of the MT architecture

by different requirements, an architectural solution was adopted based on an hybrid system model; this model, called the *wormhole* model [44], features two subsystems following different sets of assumptions (e.g., about synchrony). The middleware designed for the ALARP project is distributed on the two subsystems, and hides the hybridization to the other layers of the system.

The middleware is tightly matched to the underlying hybrid architecture and consequently to present the resilient solutions in the MT it is necessary to detail both the hybrid architecture and the middleware which executes and manages it.

In the following of this Section, we first present the various layers of the MT, and then we describe how the hybrid architecture model is applied in the MT.

2.3.1 The MT Layers

Three software layers can be identified in the MT: *(i)* application layer, *(ii)* management middleware and *(iii)* communication layer. Also ergonomic solutions, I/O devices and the hardware in general constitute an additional layer. We introduce these layers in the following of this Section (further details can be found in [3,5,37]). A high level view of the MT is in Fig. 2.2.

The MT *application layer* has three main operational modes, or states: *active* (i.e., the MT is active and normally working), *alert* (i.e., an MT enters in this state if it receives an alert event), and *warning* (i.e., the MT enters this state when it receives a *warning* event). The two macrocomponents identified in this layer are the *localization* component, in charge of locating the worker within the worksite, and the *application logic*, which contains all the functional applications required by the MT subsystem (e.g., signaling of approaching trains, health monitoring, mapping of workers in the worksite, showing information to the users).

The localization service is a highly relevant service in ALARP, as it is the mean to understand if a worker is in red zone or green zone. Similarly to *Local Area DGPS* (LADGPS, [26]), which improves the accuracy of standalone GPS by estimating errors corrupting the standalone GPS position solution and transmitting these estimates to nearby users, the ALARP approach assumes that an anchor node (point with known and precise geographic coordinates) is established on the worksite (e.g., placed a day before the working day using highly accurate instruments, or through railway maps and information available on the railway line). A GPS device placed on the anchor node can act as reference station to broadcasts a correction factor (its error with regard to the coordinates of the anchor node) to the other devices, which apply such correction to mitigate their localization errors, also relying on the information provided by electronic fences placed along the railway [37].

The *management middleware* provides facilities to the other layers, implements services for resilient MT execution, and hides to the other layers the presence of two different subsystems: wormhole and payload. The management middleware is described in details in the following Sections.

The *communication layer* provides the communication functionalities of the MTs towards the TPADs and the other MTs. The real-time communication solution adopted relies on a central coordinator which polls the various MTs, following the approach of the *Real-time Group Communication Protocol* [31]. Details on the communication layer and the real-time protocol it executes are in [6, 29].

Regarding I/O devices and ergonomics solutions, the MT contains a HMI to inform the worker on approaching trains or events that could put at risk his safety. The interface has a simple structured layout with large control buttons and a display to receive signals and fast user interactions in situations with high risks. The alert and warning signals are transmitted over the visual and acoustic human channels.

2.3.2 The MT Hybrid Architecture

Architectural hybridization is a paradigm to construct systems with wormholes (i.e., special components that present improved characteristics with respect to the remaining components of the system [44]). The *wormholes model* is adequate to describe distributed systems with uncertain or unknown attributes, such as asynchrony or failure modes. From [44], the design philosophy of the wormhole model builds on the following two guiding principles:

- First, assume that uncertainty is not ubiquitous and is not everlasting—this means that the system has some parts which are more predictable than others and tends to stabilize;
- Then, be proactive in achieving predictability—in concrete, make predictability happen at the right time and right place.

In the MT, some services are assigned more urgent trusted requirements than others. Such services are located in the wormhole of the MT: the wormhole includes services useful for applications with strict time and/or with strict security requirements. The behavior of these services must be known and predictable: to not impact reliability and safety of the whole system, such services have to be provided with certain guarantees at their interfaces.

The MT architecture is organized in two subsystems, a *wormhole* subsystem and a *payload* subsystem (Fig. 2.2). The wormhole part of the architecture provides services with different quality with respect to the payload part. In particular, it assists the execution of fault tolerant algorithms, offering a minimal set of secure and timely services. This ensures separation of concerns between the properties offered by the wormhole and the payload and between the executions on either side of the *wormhole gateway* [44] which connects them.

In what follows, we introduce the services offered by the middleware, their separation on wormhole and payload and related motivations.

The wormhole is intended to execute on a very basic hardware, that ease monitoring and assessing its behaviour. The payload executes on a different hardware than the wormhole. The payload requires several hardware devices, and especially I/O devices to interact with the workers and network devices to communicate to the TPADs and the other MTs.

Two different part of the middleware executes on the wormhole and on the payload, and communicate through the wormhole gateway. Also, the middleware is the only software layer, amongst those previously enlisted, which executes on the wormhole.

In the wormhole, simple but critical services reside. The services identified are the following middleware services (Fig. 2.2): *(i)* time-related services (the resilient clock *Reliable and Self-Aware Clock*, that estimates time uncertainty and provides trusted time values, and a clock synchronization mechanism), and *(ii)* a privacy and authentication service (it stores the private key of the worker, which allows signing the messages that are sent to provide authentication and non-repudiation). Such services are located in the wormhole because they require stronger (with respect to the rest of the system) assumptions on timeliness and security. Payload processes uses the wormhole services through the wormhole gateway. The payload processes do not need to know how wormholes are implemented, and viceversa. This model ensures a clear separation of concerns between the properties offered by wormholes and their construction, and between the executions on either side of the wormhole gateway.

Regarding services executing on the payload, these are mainly monitoring and fault tolerance mechanisms that provide the required SIL of the MT, and are enlisted in what follows. The *execution monitor* supervises the execution of the various software components: it handles the activation and deactivation of threads and functions, and it monitors the execution of MT software. The *detection and testing* performs software tests and checks on the MT. The *Quality of Service monitor* (*QoS monitor*) provide to the other layers information on *(i)* the battery power status (it also notifies when battery power is low), *(ii)* the quality of local clock (synchronization uncertainty), achieved querying the wormhole service R&SAClock, and *(iii)* the localization uncertainty. The *interface manager* allows communicating with the communication layer and the application layer, and provides interfaces to the hardware drivers. The *security manager* manages public keys of the other components of the system and executes algorithms for message deciphering and signature verification.

2.4 State Machine of the MT Middleware

The MT interacts with the other entities in its environment exchanging information. The overall behaviour of the middleware is performed by a set of global states that can be expressed by the state machine depicted in Fig. 2.3 (note that two other state machines are identified for the application and communication layers, but are not explored in this paper; details can be found in [3, 5, 6]). Table 2.1 defines the MT behaviour when state changes occur.

The MT management middleware has five operational modes, or *states*. In *Start up* state, initialization procedure and all devices are tested. If a fault happens, a transition to the Safe state is performed. Otherwise, it means that the MT is ready to start, and consequently Nominal state or Low Power states are entered.

In *Low Power* state, the MT has low battery power autonomy and consequently it provides a subset of its functionalities in order to save batteries. For example, it reduces to a minimum the number of messages the MT exchanges. In fact, the messages exchanged by the MT are of three criticality levels, that are *high* for safety-critical messages (e.g., alert messages), *medium* for messages which are critical

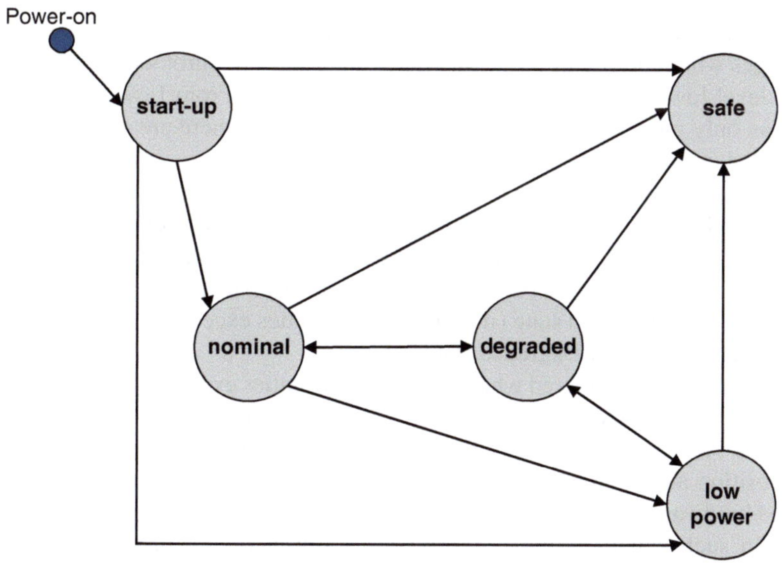

Fig. 2.3 State machine of the MT middleware

Table 2.1 Actions at state change

From state	To state Start up	Nominal	Low power	Degraded	Safe
Start up	–	Establish connection with TPAD(s) and start nominal operations	Establish connection with TPAD(s), signal low power to the worker and start operations for low power state (reduce messages, display only relevant events)	–	Signal to the worker, broadcast message to other workers (if able to broadcast)
Nominal	–	–	Signal to the worker	Signal to the worker	
Low power	–	–	–	–	
Degraded	–	Signal to the worker and start nominal operations	Signal to the worker and start operations for low-power state	–	
Safe	–	–	–	–	

for system availability (e.g., messages to signal the alert termination), and *low* for messages which do not impact availability and safety requirements. In Low Power state, level low messages are not exchanged anymore. From Low Power state, an MT can only move to Safe state or being switched off (there are no procedures for online battery replacing).

In *Nominal state*, the MT is fully operative. This state can be entered from Start up state or Degraded state. In Nominal state, periodic testing activities are executed and diagnostic functionalities are available. From Nominal state, the MT can enter Safe state (if an error is detected), Low Power state (if battery power is under a given threshold) and Degraded state (if system uncertainties exceed a given threshold, see below the definition of Degraded state).

Degraded state is entered when system uncertainties exceed a given threshold. In the case of the MT requirements, examples of uncertainties considered are temporal and spatial uncertainties. That is, Degraded state is entered when uncertainty on time and position are smaller than the allowed thresholds. Also, failures of non safety-related hardware (for example display, or audio) may cause a transition to degraded state. In such situations the MT provides a subset of its functionalities, due to lack of information (for example poor clock synchronization, or large uncertainties in localization). Whenever such uncertainties are reduced the MT can return to nominal state.

Safe State is entered when a malfunction is detected. When the MT middleware moves to *safe state*, the MT signals to the worker that it is not able to properly function anymore and halts its services. When such state is reached the worker has to perform a failover procedure which consists in trying to manually reboot the MT or change it with a spare one.

2.5 The Wormhole and the Wormhole Gateway

The wormhole is devoted to provide trusted middleware services. These services have been located in the wormhole because they require stronger (with respect to the rest of the system) assumptions regarding time and security. We start the discussion on the wormhole first presenting its hardware, and then the middleware software components which execute on it. Finally, we describe the wormhole gateway.

2.5.1 Wormhole Hardware

The wormhole is intended to execute on a very simple hardware (a CPU, a RAM, a Flash component and an external memory, see Fig. 2.4), that ease monitoring and assessing its behavior, as it only needs to manage private information and reliable timekeeping. We can note that there is no watchdog timer: if the wormhole halts or crashes, the payload can notice its failure through its own timers (though in such a case a violation of the real-time requirements must be taken into account). The external memory of the wormhole is a *pluggable memory* (preferably a *smart card*) which contains sensible information of the worker, as its personal data and the private key that are used to authenticate messages. To protect the information and trust of the worker in the MT, the smart cart brings the following main benefits: *(i) inbuilt security*, because smart cards can be tamper-resistant and

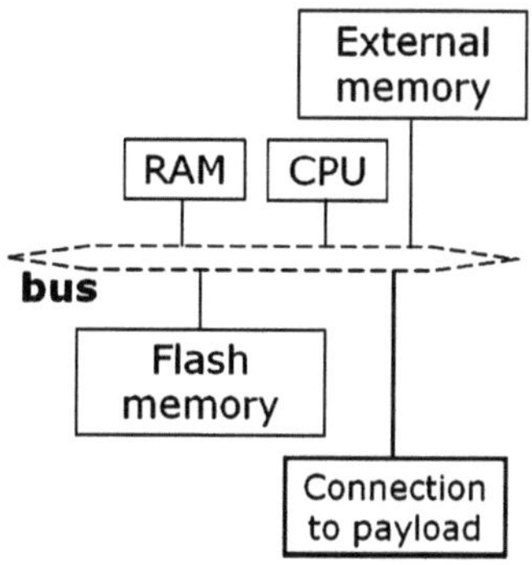

Fig. 2.4 View of the wormhole hardware architecture

protect information that is stored on them, and *(ii) usability and portability*: in fact, smart cards are portable, small, and easy to use.

2.5.2 *Wormhole Software*

We detail the software modules which are executing on the wormhole.

2.5.2.1 Authentication

The *authentication* component is devoted to manage part of the authentication procedure for key storage and data ciphering. It is in charge of importing and using the private key of the worker (which is contained in the smart card) to sign messages according to requests from the payload. The authentication component exposes to the payload the method *sign(message)*, which calculate the digital signature of a message using encryption with *Secure Hash Standard* (SHA) [32] and *Rivest, Shamire, Adelman* (RSA) [34]. When this method is invoked, it calculates the digest of the message with the SHA algorithm, and then it applies the RSA encryption to the digest originally computed.

Details on the authentication procedure adopted in ALARP are reported in Sect. 2.6, when presenting the payload services; in fact the authentication service in the wormhole just signs messages provided by the payload using the private key stored in the smart card; all the remaining procedures (interactions for the initially agreement on the set of MTs and their matched keys, keys exchange, storage of the public keys of the various MTs, and signature verification) are demanded to the payload service *security manager*.

2.5.2.2 Synchronization Mechanism

Adaptive and distributed systems as the set of the MTs cooperating in the ALARP system require clocks to be synchronized to an external global timeline in order to correctly execute their services [35, 45]. A good clock synchronization is a crucial requirement, as failing to fulfill it can have a severe effect on the performance or even the safety of a system [10].

The *synchronization mechanism* adopted for the MT has the role to discipline the software clock of the wormhole. As a global system view shared by the MTs is of utmost importance, a synchronization mechanism is needed (together with a component to estimate the uncertainty in the synchronization for each node, see following section).

The MT uses an external time reference for clock synchronization. The most widespread timescale is the *Coordinated Universal Time* (UTC) [27], but in UTC a leap second is periodically inserted: the UTC time-growth function presents a discontinuity due to the leap second insertion action. Instead the *Temps Atomique International* (TAI) [27] has the same growth rate of UTC, with no discontinuity: this simplifies the development of time-related algorithms. Converting UTC data to TAI data and viceversa is low-effort [27], so we can use synchronization mechanisms based on UTC without significant increase in computational or development costs [8].

Given the external time reference, an external clock synchronization mechanisms is needed. The mechanism chosen is the *Network Time Protocol* (NTP) [30] as synchronization mechanism (note that NTP is based on UTC, so a conversion function is needed).

2.5.2.3 Reliable and Self-Aware Clock

While having their clock synchronized to an external one, a number of systems still need to know their offset from the global time, in order to correctly execute their services. However, due to its stochastic nature, and to its strong dependence on unpredictable environmental changes and/or changes in the system dynamics and faults, the offset is usually quite hard to compute. Though many synchronization mechanisms perform an offset estimation, they offer limited guarantee on the closeness of this value to the actual offset, and usually express this limit in terms of a worst case bound. In fact, unfortunately, the offset is typically a variable factor very hard to compute due to changes in system dynamics, faults, or environmental changes [21, 43]. So, either the offset estimation is hardly trustable or—when expressed as a worst case bound—its value comes out to be very large and far from typical execution scenarios and, consequently, of little use in practice [10].

The notion of uncertainty of the time measurement as used in metrology [25] can provide an estimation of synchronization quality that in practice can result more useful than accuracy and offset. To this reason, we define *synchronization uncertainty* [8] as an adaptive and conservative estimation of the distance of local clock from global time.

The *Reliable and Self-Aware Clock* (R&SAClock) [8, 9] is a software clock for resilient time information that computes synchronization uncertainty and provides to its users both current time and current synchronization uncertainty. R&SAClock is not a synchronization mechanism, but it acts as a software local clock that exploits services and data provided by any chosen synchronization mechanism for external clock synchronization. The R&SAClock is designed to work with any kind of external synchronization mechanisms and provides reliable information even in case the synchronization mechanisms in use are not working properly, or inadequate conditions for synchronization happen (e.g., the network becomes unavailable). In the context of ALARP, the R&SAClock is matched to the NTP protocol. In what follows, we present basics on time and clocks (see also Fig. 2.5), necessary to introduce the R&SAClock, a brief description of the key algorithm implemented, and a sample trace of results collected. Extensive discussions on the insights of the R&SAClock and simulation and experimental results can be found in [8–10].

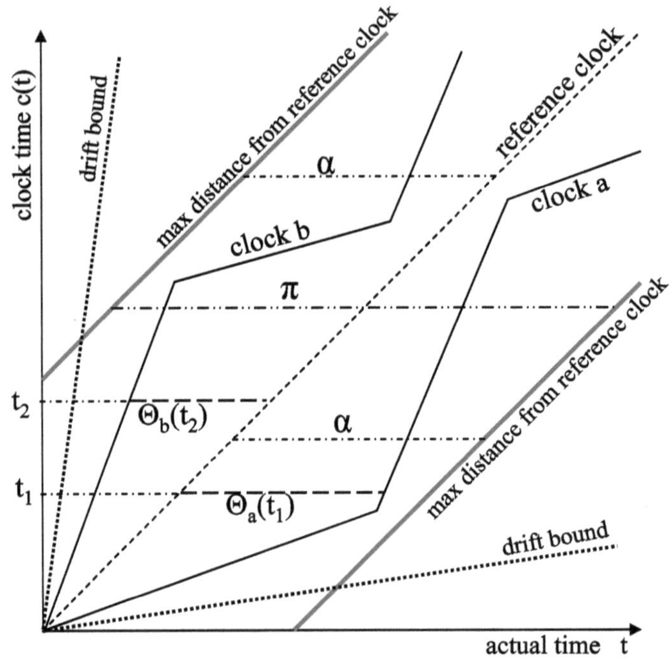

Fig. 2.5 Basics on time and clocks

Basic Notions of Time and Clocks. Let us consider a distributed system composed of a set of nodes. We define *global time* as the unique time view shared by the nodes of the system, *reference clock* as the clock that always holds the global time, and *reference node* as the node that owns the reference clock. Given a local clock c and any *time instant* t, we define $c(t)$ as the *time value* read by local clock c at time t. The behavior of a local clock c is characterized by the quantities *offset, accuracy* and *drift*. The offset $\Theta_c(t) = t - c(t)$ is the actual distance of local clock c of the node n from the global time at time t [30]. This distance may vary through time. Accuracy A_c is an upper bound of the offset [45]; accuracy is often adopted in the definition of system requirements and therefore targeted by clock synchronization mechanisms. Drift $\rho_c(t)$ describes the rate of deviation of a local clock c at time t from global time [45].

Despite their theoretical importance, accuracy and offset are usually of practical little use for systems. Synchronization mechanisms typically compute an *estimated offset* $\tilde{\Theta}_c(t)$ and an estimated drift $\tilde{\rho}_c(t)$, without offering guarantees and only at synchronization instants. We define the *synchronization uncertainty*, or simply uncertainty, $U_c(t)$ as an adaptive and conservative evaluation of offset $\Theta_c(t)$ at any time t; uncertainty is such that $A_c \geq U_c(t) \geq \Theta_c(t)| \geq 0$.

Specification of the R&SAClock. R&SAClock is a software clock that provides to users (e.g., a process of the MT) both the time value and the synchronization

Table 2.2 Main SPS quantities and parameters

Symbol	Definition
t_0	Time in which the most recent synchronization is performed
$\tilde{\Theta}(t_0)$	Estimated offset at time t_0
$\tilde{\rho}(t_0)$	Estimated drift at time t_0
M, m	Maximum and current number of (most recent) samples of the estimated drift collected ($0 < m \leq M$)
N, n	Maximum and current number of (most recent) samples of the estimated offset collected ($0 < n \leq N$)
p_{ds}	Probability that the population variance of the estimated drift is smaller than a safe bound on such variance
p_{dv}	A safe bound of the drift variation since t_0 is computed with probability p_{dv}
$p_{ds} \circ p_{dv}$	The joint probability of these two values represents the coverage of the prediction function
p_{os}	Probability that the population variance of the estimated offset is smaller than a safe bound on the variance
p_{ov}	A safe bound of the offset at t_0 is computed with probability p_{ov}
$p_{os} \circ p_{ov}$	The joint probability of these two values represents the coverage of the safety margin function

uncertainty associated to the time value. When a user asks the current time to R&SAClock (by invoking function *getTime*), R&SAClock provides an enriched time value [*likelyTime*, *minTime*, *maxTime*]. *LikelyTime* is the time value computed reading the local clock (i.e., $c(t)$). *MinTime* and *maxTime* are based on the synchronization uncertainty provided by the internal mechanisms of R&SAClock.

From the perspective of an MT process interested in timeliness of services, the main expectations from the R&SAClock are the following: *(i)* a request for the time value should be satisfied quickly, and *(ii)* the enriched time value should include the true time. These can be more formally expressed as:

REQ1. The service response time provided by R&SAClock is bounded: there exists a maximum reply time Δ_{RT} from a *getTime* request made by a user to the delivery of the enriched time value (the probability that the *getTime* is not provided within Δ_{RT} is negligible).

REQ2. For any *minTime* and *maxTime* in any enriched time value generated at time t, it must be $minTime \leq t \leq maxTime$ with a coverage Δ_{CV} (by coverage we mean the probability that this equation is true). In other words, given *likelyTime* $= c(t)$, the true time t must be guaranteed within the interval [*minTime*, *maxTime*] with a coverage Δ_{CV}.

The Statistical Predictor and Safety Margin (SPS). We briefly describe the SPS algorithm for a local software clock c that is disciplined by an external clock synchronization mechanism. A complete description of the algorithm can be found in [9]. To ease the readability of the notation, the subscript c is omitted from the expressions presented in the rest of this paper. In Table 2.2 the main quantities involved in the SPS are shown and explained.

We assume t_0 the time in which the most recent synchronization is performed: at time t_0 the synchronization mechanism computes the estimated offset $\tilde{\Theta}(t_0)$ and possibly the estimated drift $\tilde{\rho}(t_0)$ (if not provided by the synchronization mechanism, it can be easily computed by R&SAClock itself).

SPS computes the uncertainty at a time t with a coverage, intended as the probability that $A \geq U(t) \geq \Theta(t)| \geq 0$ holds. The computed uncertainty is composed by three quantities: *(i)* the *estimated offset*, *(ii)* the output of a *predictor* function P and *(iii)* the output of a *safety margin* function SM. The computation of synchronization uncertainty requires a right uncertainty $U_r(t)$ and a left uncertainty $U_l(t)$: consequently SPS has a right predictor and a right safety margin for right uncertainty, and a left predictor and a left safety margin for left uncertainty. The output of the SPS at $t \geq t_0$ is constituted by the two values:

$$U_r(t) = \max(0, \tilde{\Theta}(t_0)) + P_r(t) + SM_r(t_0) \tag{2.1}$$

$$U_l(t) = \min(0, \tilde{\Theta}(t_0)) + P_l(t) + SM_l(t_0) \tag{2.2}$$

The *predictor* functions (left and right) predict the behavior of the oscillator and continuously provide bounds (lower and upper) which constitute a safe (pessimistic) estimation of the oscillator drift and consequently a bound on the offset. The oscillator drift is modeled with the random walk frequency noise model, one of the five canonical models used to model oscillators (the power-law models [7]), that we considered as appropriate and used.

The *safety margin* functions (left and right) aim at compensating possible errors in the prediction and/or in the estimation of the offset.

The setup parameters used by SPS (detailed in Table 2.2) are: the four probabilities p_{ds}, p_{dv}, p_{os}, p_{ov} and the values M and N. The coverage and the performance (how much the synchronization uncertainty is effectively tight to the estimated offset) achieved by SPS depend on these parameters.

R&SAClock into practice. To apply the R&SAClock to the ALARP scenario, a tuning activity of its parameters is needed. An experimental evaluation of R&SAClock using SPS is discussed in [10]. In such evaluation the enriched time values provided by the R&SAClock under varying synchronization conditions are confronted with the time instants provided by an high quality clock providing a very good approximation of the reference time.

A sample result of the experimental validation activity performed on the R&SA-Clock software executing on a common PC is reported in Fig. 2.6, which shows the results for $p_{ds} = p_{dv} = p_{os} = p_{ov} = 1\text{–}10^{-6}$, the workload being 1 requests/seconds, and the memory depth $M = N$ being 40 samples. In this run of the duration of two hours, the enriched time values provided by the R&SAClock are compared with a highly accurate time reference (a clock tightly synchronized with microsecond accuracy, that we use as reference clock). The interval (*minTime*, *maxTime*) provided by R&SAClock is narrow, and always succeed in containing the true time (the values provided by the perfect clock).

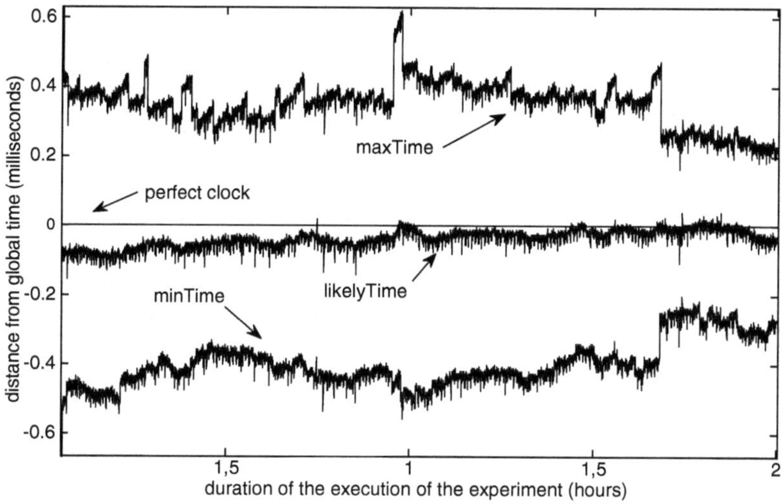

Fig. 2.6 Synchronization uncertainty for coverage level $p = 0.999999$, $M = N = 40$

2.5.3 The Wormhole Gateway

Payload processes uses the wormhole services through the *wormhole gateway*; for payload processes the properties offered by any wormhole are defined at such wormhole gateway. The payload processes do not need to know how wormholes are implemented, and viceversa. This ensures a clear separation of concerns between the properties offered by wormholes and their construction, and between the executions on either side of the wormhole gateway.

In ALARP, the wormhole gateway is implemented as a singleton, and the communication between wormhole and payload is based on sockets. The methods exposed by the wormhole gateway are the only way for the payload to use the wormhole features (and viceversa). The main methods exposed are *(i)* an interface toward the payload to command hardware test, which the wormhole can use to invoke tests performed and coordinated by the payload, *(ii)* public methods to be invoked by the payload to use the wormhole features (to get the enriched time value provided by the R&SAClock, and to sign messages), *(iii)* an interface towards the payload to send messages for the execution of the clock synchronization algorithm.

2.6 The Payload Middleware

The payload subsystem contains the application layer, (part of) the management middleware, and the communication layer. We first present the payload hardware, then its middleware software components which execute on it.

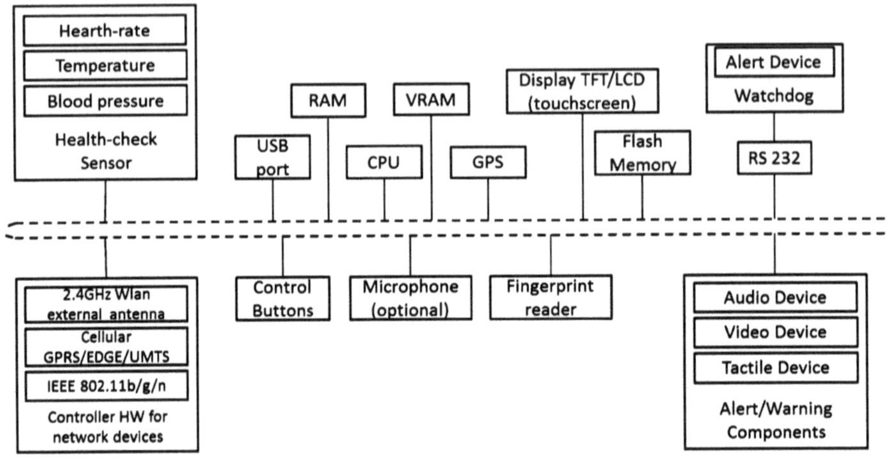

Fig. 2.7 Hardware components of the payload

2.6.1 Hardware

As discussed for the wormhole, the design of the payload middleware strictly depends on the hardware selected. In fact, the payload middleware contains mechanism to manage and query the hardware devices, including mechanisms for hardware fault tolerance. Consequently, although this paper focuses on the middleware, a brief description of the required hardware is necessary. The hardware components of the payload are represented in Fig. 2.7.

A bus allows the interactions between the *Central Processing Unit* (CPU) and the other devices by means of a bus controller. The data storage is based on a flash memory, or *Solid State Disk* (SSD), that contains the payload software, configuration data, graphics icons, audio samples, etc. Note that (traditional) hard disks are avoided because particularly subject to failure in case of relevant vibrations. Common components as a RAM, a videopage to display graphics to the worker, and an USB port (e.g., for uploading new software versions) are also necessary.

Health-check sensors/devices for biometric monitoring are necessary to measure heart rate, blood pressure and temperature, as well as alerting/warning components for different human sensors (minimum two channels are used: visual and audio). Also, a GPS device for positioning and clock synchronization is present.

The main hardware interfaces for communication are the following: a single antenna WiFi 802.11b/g/n, a cellular module for GPRS/EDGE/UMTS communication, an external 2.4GHz WLAN antenna and a standard controller for communication logic.

To interface with the worker, identified components are a TFT-LCD display (optionally touchscreen), and control-buttons. In case voice communication is added

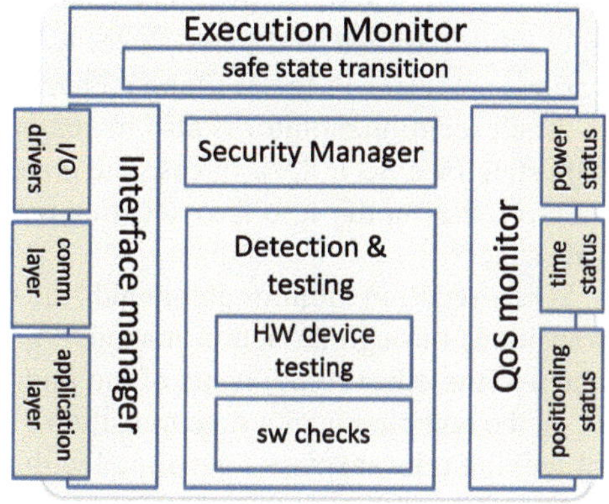

Fig. 2.8 Main software components of the payload middleware

as an additional feature, a microphone is required. Similarly, a fingerprint reader may be added to ease the authentication process of the MT owner.

To signal the MT incorrect behavior to the worker, a signaling system which is alternative to the one described above is required. This signaling system consists in a separate watchdog hardware which detects the incorrect behavior of the MT, and a separate signaling channel activated by the watchdog to notify the worker. For example, this is needed in case the CPU presents a halt failure, and consequently is unable to sends outputs to workers. Note that alternative signaling notifications as switching off displays, usually applied in the railway field for safety-critical notifications, are not applicable due to ergonomics issues. The watchdog consists in a very basic hardware, separated by the wormhole and the payload, with its own battery pack and output device to the worker. The only functionality of the watchdog is to command an alarm to the worker whenever the watchdog is not refreshed by the payload middleware software.

2.6.2 *Middleware Services*

The software architecture of the *management middleware* is depicted in Fig. 2.8. Its main components are the execution monitor, the security manager, the interface manager, the detection and testing component, and the QoS monitor. We detail these components in what follow.

2.6.2.1 Execution Monitor

The *execution monitor* supervises the execution of the various software components. It handles the activation and deactivation (killing) of threads and functions; e.g. at state change it commands the killing of processes and services that are no more needed, and the activation of new processes or services that will execute the operations required in the new state. Additionally, it monitors the execution of the various processes of the MT, and it checks that real-time deadlines are met.

The execution monitor is also in charge to command transition of the MT to the safe state (see Sect. 2.4). In case the error detection mechanisms of the MT detect an error, they notify it to the execution monitor which commands a transition to the safe state.

The execution monitor commands the execution of actions (for example, a state transition) through an action manager, and generates events which are communicated to the other components of the middleware and to the other layers of the MT (e.g., the completion of a state transition). Finally, the execution monitor is in charge of logging relevant events generated within the middleware.

2.6.2.2 Detection and Testing

The *detection and testing* component performs software tests and checks on the MT, to detect permanent and transient failures in the hardware, and systematic failures in the software (e.g., software bugs); that is, this module is in charge of the error detection mechanisms in the middleware.

Following the approach proposed in [12], active *permanent faults* and *random transient faults* in the hardware components, and residual systematic software faults (e.g., bugs that are not detected during testing activities), are detected on the basis of their effects on software execution. Instead, periodic testing is performed to detect latent *random permanent faults* in the hardware components.

The component detection and testing is divided in two components: *hw device testing* and *sw checks*. The module *hw device testing* performs online and offline tests of the hardware devices of the MT, to activate dormant hardware faults and detect the related errors. In particular, RAM and flash memory are tested by March test algorithms [42] (similar tests are applied to the Video Page), the CPU is checked by specific test suite to address faults in its major components as registers, arithmetic and logic unit, and instruction decoding [39], and testing of the peripheral devices (LCD, audio device, etc.) relies on periodic or start up tests procedures which require interactions with the worker (e.g., to verify that the LCD is actually blinking, or the audio device is regularly working).

The module *sw checks* contains software mechanisms to detect active permanent faults and random transient faults in the hardware components, and residual systematic errors (e.g., software bugs) on the basis of their effects on software execution. Following the classification in [38], the set of mechanisms implemented is: *(i) data acceptance/credibility checks* (which are efficient in the case of data processing functions, but only if it is possible to assign acceptance criteria to intermediate or output data), *(ii) control flow monitoring* techniques (efficient in control-related functions; however, its coverage is limited, as the result of the computation could be erroneous even if the control flow itself was correct, while data acceptance checks could also detect if the computation was affected by a control flow error), *(iii) timeout checking* (applied whenever timing constraints can be defined [28]) and *(iv) periodic wormhole service checking* (i.e., services offered by the wormhole can be periodically queried to verify their proper behavior).

2.6.2.3 QoS Monitor

This component monitors the uncertainties that are part of the MT. Information provided by the QoS monitor enables the transitions *to* and *from* the Low Power state and the Degraded state. The *QoS monitor* monitors:

1. the *battery power* of the MT, to provide to the other layers information on the battery power status and to notify when battery power is low;
2. *quality of clock* (synchronization uncertainty), monitored by intercepting answers provided by the R&SAClock to the services that use it (the QoS monitor does not invoke the R&SAClock directly, but only intercepts messages, in order to limit as much as possible the number of messages exchanged with the wormhole);
3. *quality of localization*, to provide information on localization uncertainty (possible uncertainties in the computation of the position of workers).

Point *(1)* of the above list requires to monitor the battery level of the MT, and can be realized using functions typically provided by the operating system.

Point *(2)* instead is realized through the service offered by the R&SAClock. The R&SAClock and synchronization uncertainty is described in details in Sect. 2.5.2.3; we do not further explore such discussion here.

Instead, point *(3)* addresses another relevant source of uncertainty that is very significant in the MT context: localization uncertainty. The position computed by the localization module, which is in the application layer of the MT, is most likely affected by a measurement error thus introducing uncertainty in deciding if the worker is in red or green zone; the objective of the QoS monitor is to be aware of the kind of error sources, and the impact of measurement errors on position estimation. This allows in the end to provide an estimation of position uncertainty. To build and configure the QoS monitor in ALARP, an investigation on position uncertainty, localization errors and their sources is required.

From a classification of the sources of localization errors [26], those which have significant effect in the ALARP scenario are singled out.

The analysis on the sources of errors and their contribution starts from the following considerations: *(i)* a typical railway worksite, in which the workers (the MTs) need to be localized, is a limited operation area (maximum 700 m length), *(ii)* workers typically moves on foot (slow movement speed), *(iii)* the worksite can be located in place possibly surrounded by foliage, in canyon, or near buildings (i.e., the MTs may suffer limited visibility of GPS satellites), and finally *(iv)* the GPS devices used should be low cost and COTS.

Consequently, in the ALARP scenario, all GPS devices are the same model and the same antenna sensitivity; moreover, they are very close to each other (up to

700 m). This implies that in most of the situations they will share the same satellites visibility: consequently, they will share very similar ephemeral and satellite clock errors (i.e., errors in the synchronization of the satellites clock and in the definition of their orbits [26]). Also, receivers are sufficiently close to share similar ionospheric and tropospheric errors (i.e., errors that are causing by the transition of the signal in the ionosphere and troposphere). Finally, the devices are of the same model and consequently their design will influence the position error in a similar manner.

It is reasonable to expect that in our scenario errors due to satellite clock and ephemeral satellite orbits (when receivers use the same satellite set), ionospheric and tropospheric signal perturbation and delays, and the receiver's design can be considered as systematic. At the same time, we can expect that errors due the receiver's thermal noise and external interferences exhibit negligible variations from a receiver to the others. Conversely, due to its nature, multipath (reflections of the signal on the ground or nearby objects [26]) affects the localization error randomly.

To verify the behavior (and error) of the COTS GPS devices identified for the ALARP prototype and to assess the error correlation of such devices in different worksite-like scenarios, an experimental campaign has been planned and executed [11]. The work [11] provides details on the experiments motivations and planning, and results achieved through a first test session to evaluate two popular GPS devices, the Globalsat ND100S [23] and the Garmin LVC 18 [22], using a precise positioning system (the Trimble R7 [40] and R8 [41]) as reference.

To summarize the achieved results, the following observations are reported. First, possible systematic contributions to the localization error can be identified for the Globalsat devices when doing stationary measurements; for a short period of time, in similar environmental conditions and satellite visibility, similar localization errors are measured. On the contrary, the behavior of the Globalsat devices looks more unstable during motion measurements, because they showed irregular sampling periods in which data are not continuously refreshed. Second, during motion measurements, encouraging results were instead given by the Garmin LVC 18, whose estimated positions always fluctuated around the true values.

2.6.2.4 Security Manager

Although the MTs communicate using a non-encrypted wireless links, digitally signed messages are required for communication amongst MTs (and to the TPADs), to provide authentication, integrity and non-repudiation [38]. An asymmetric authentication mechanism is provided where each MT has a couple of public-private keys. In this mechanism, the authentication service of the wormhole is only in charge of storing the private key, and signing (whenever requested by the payload) messages that are sent to other system components. Instead, the security manager on the payload is in charge of collecting the public keys of the other system components, and verifying messages using such public keys. We note that through this separation of the security service in two parts, we locate the service which manages critical data (the private key) on the wormhole, since it is expected to satisfy more stringent requirements in terms of security, and the service which manages public data (the public keys) on the payload. The *security manager* has the following functionalities:

- It invokes the authentication service of the wormhole to sign message (before they are sent to other MTs);
- It stores the public keys of the other devices (MTs and TPADs) that are part of the ALARP system. Such public keys are used to verify the signature of messages received.

2.6.2.5 Interface Management

Finally, the component *Interface management* allows communicating with the communication layer and the application layer, and allows interfacing with the hardware drivers

2.7 Towards the MT Prototype

As the ALARP project is ongoing at the moment this paper is written, the implementation of the middleware and of the overall MT prototype is still not complete. In this Section we briefly summarize the main envisioned solutions for its implementation (in the ALARP project, different prototypes are planned, to be prepared at different stages of the project [1]).

The approach currently explored is based on the usage of generic laptops, which offers flexible software platforms for implementing specific applications and connect external devices. To implement the hybrid architecture model using such devices, an approach based on virtualization platform is selected. The approach for these prototypes consists in instantiating two virtual machines running on the same laptop: one virtual machine acts as the wormhole, the other as the payload. This provides isolation of wormhole and payload to software faults, but it does not guarantee full hardware independency.

To achieve at least some level of hardware/operating system independency, the Xen 4.0 [46] Linux Virtualization platform was selected: Xen allows allocating different virtual machines to different CPU cores. The Xen hypervisor, that can be considered a modified version of a Linux operating system, interfaces the hardware layer to the various guest systems (also called *domains*), managing the resources available and sharing the hardware with different operating systems installed in the domains.

To realize the physical distinction between payload and wormhole, instead, still avoiding the costs of using fully-redundant systems, the envisioned solution for a second prototype is based on multicore processors, where wormhole and payload are executing on different processors. This guarantees some level of independency

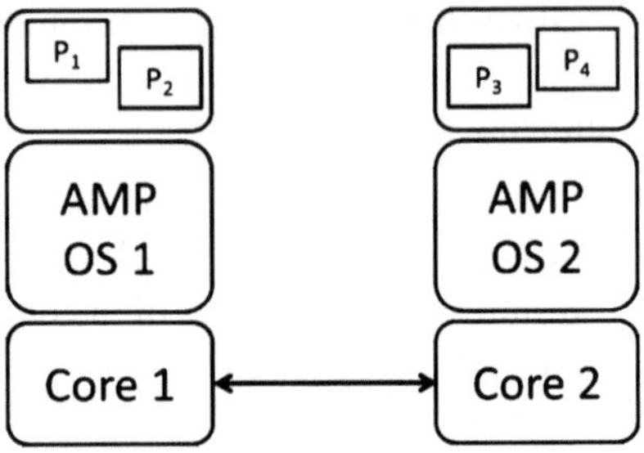

Fig. 2.9 AMP technology on dualcore processors

(see requirements for SIL 2 systems reported in Sect. 2.2) of the wormhole and payload, still having a hardware solution which is cheap, ready-to-use and easily available on market.

Further hardware independency between the wormhole and payload with COTS components can be achieved using approaches based on *Asymmetric Multi Processing* (AMP), to execute separated software and *Operating Systems* (OSs) on two (almost completely) isolated hardware platforms which communicate one with the other.

The AMP technique and the dual core architecture can be applied to prototype the MT as follow (see Fig. 2.9): the payload processes (p_1 and p_2 in figure) run on OS1 which in turn runs on core 1, while the wormhole processes (p_3 and p_4 in figure) run on OS2 which in turn runs on core 2. Inter-core communication is used in order to exchange information between wormhole and payload processes.

2.8 Conclusions

Executing safety critical and real-time services in open and harsh environment through small and portable devices is a hard task, which requires solutions for resilience and self-adaptiveness of services to changes in the surrounding operating conditions and consequently to the related uncertainties.

In this paper we presented a resilient solution for wearable mobile devices, the Mobile Terminals (MTs), which are developed as part of the ALARP project. The goal of the ALARP project is the study, design and development of a safety-critical and real-time Automatic Track Warning System for railway trackside workers. Primary goal is to recall the attention of a workgroup operating on a railway worksite, through the MTs, about the presence of trains approaching the worksite.

The hybrid architecture system model adopted in the MT design is the wormhole model that allows separating the architecture of the MT in two parts, a wormhole and a payload part, characterized by different requirements in terms of timeliness and security.

The management middleware of the MT executes on the two parts of the hybrid architecture, providing the required fault tolerant layer and hiding to the other layers the subdivision of the architecture in two parts.

A prototype of the MT shall be developed by the end of the project; the envisioned approaches to implement the hybrid system (the wormhole and payload) are based on cheap, COTS components exploiting virtualization and multicoring for the execution of the different tasks on separate cores.

Acknowledgements This work has been partially supported by the European Project FP7-234088 ALARP (A railway automatic track warning system based on distributed personal mobile terminals). The authors would like to thank all the partners and the collaborators involved in the design and implementation of the MT for their precious contributions, and in particular Alessandro Daidone, Carlo Dambra, Luca Ghelardoni, Florjan Gogaj, Jesper Grønbæk, Davide Iacono, Boris Malinowsky, Michael Schultheis, Hans-Peter Schwefel, Andrea Seminatore, Iuri Tempestini and Jurij Wakula, Joao Figueiras.

References

1. ALARP – A railway automatic track warning system based on distributed personal mobile terminals – FP7-IST-2010-234088
2. ALARP Consortium (2010) D1.3- Overall Architecture Design. EU FP7 234088 project ALARP, Deliverable 1.3
3. ALARP Consortium (2010) D3.1- MT Architecture Design. EU FP7 234088 project ALARP, Deliverable 3.1
4. ALARP Consortium (2010) D1.2- Requirements specifications. EU FP7 234088 project ALARP, Deliverable D1.2
5. ALARP Consortium (2011) D3.2- Preliminary MT Design. EU FP7 234088 project ALARP, Deliverable 3.2
6. ALARP Consortium (2011) D2.2- Preliminary Wireless Communication Solution. EU FP7 234088 project ALARP, Deliverable D2.2
7. Barnes JA et al (1970) Characterization of frequency stability. IEEE Trans Instrum Meas 20:105–120
8. Bondavalli A, Ceccarelli A, Falai L (2008) Assuring resilient time synchronization. In: IEEE symposium on reliable distributed systems (SRDS'08). IEEE Computer Society 2008, Los Alamitos, CA, USA, pp 3–12
9. Bondavalli A, Brancati F, Ceccarelli A (2009) Safe estimation of time uncertainty of local clocks, International Symposium on Precision Clock Synchronization for Measurement, Control and Communication. ISPCS 2009. pp 1–6, 12–16
10. Bondavalli A, Brancati F, Ceccarelli A, Vadursi M (2010) Experimental validation of a synchronization uncertainty-aware software clock. In: 29th IEEE symposium on reliable distributed systems (SRDS2010). IEEE Computer Society 2010, Washington DC, USA, pp 245–254
11. Bondavalli A, Ceccarelli A, Gogaj F, Seminatore A, Vadursi M (2011) Localization errors of low-cost GPS devices in railway worksite-like scenarios. In: IEEE international workshop on measurements and networking (M&N 2011), Capri, Italy, pp 6–11
12. Ceccarelli A, Majzik I, Iovino D, Caneschi F, Pinter G, Bondavalli A (2008) A resilient SIL 2 driver machine interface for train control systems. In: Proceedings of the 2008 third international conference on dependability of computer systems (DepCoS-RELCOMEX'08). IEEE Computer Society, Washington, DC, pp 365–374
13. CENELEC (2000) EN 50129 – Railways applications – Communications, signalling and processing systems - Safety related electronic systems for signaling
14. CENELEC (2001) EN 50128 – Railways applications – Communications, signalling and processing systems – Software for railways control and protection system

15. CENELEC (2001) EN 50159-1 – Railway applications – Communication, signalling and processing systems – Part 1- Safety related communication in closed transmission systems
16. CENELEC (2001) EN 50159-2 – Railway applications – Communication, signalling and processing systems – Part 2- Safety related communication in open transmission systems
17. CENELEC (2006) EN 50126 – Railway applications – The specification and demonstration of Reliability, Availability, Maintainability and Safety (RAMS) – Part 1: Basic requirements and generic process
18. Dana PH (1997) Global positioning system (GPS) time dissemination for real-time applications. Real-Time Syst 12(1):9–40
19. Druidi D (2007) Railroad-related work injury fatalities. Monthly labor review
20. Fascicule 63- Mesures de securitée et de sante lors de l'execution de travaux geres par Infrabel. Infrabel
21. Freris NM, Kumar PR (2007) Fundamental limits on synchronization of affine clocks in networks. In: 46th IEEE conference on decision and control. IEEE, pp 921–926
22. Garmin 18 LVC datasheet. http://static.garmincdn.com/pumac/425_TechnicalSpecification.pdf. Accessed 22 March 2012
23. Globalsat ND 100S datasheet. http://www.usglobalsat.com/store/download/590/nd100s_ds_ug.pdf. Accessed 22 March 2012
24. GUV-R 2150 (1998) Sicherungsmaßnahmen bei Arbeiten im Gleisbereich von Eisenbahnen. Deutsche Gesetzliche Unfallversicherung (DGUV)
25. JCGM (2008) Guide to the expression of uncertainty in measurement, 3rd edn. JCGM stands for Joint Committee for Guides in Metrology
26. Kaplan ED, Hegarty CJ (2006) Understanding GPS: principles and applications, 2nd edn. Artech House, Boston
27. Levine J, Mills D (2000) Using the network time protocol (ntp) to transmit international atomic time (tai). In: 32nd annual precise time and time interval systems and applications meeting, Reston, pp 431–440
28. Majzik I, Hohl W, Pataricza A, Sieh V (1996) Multiprocessor checking using watchdog processors. Int J Comput Syst Sci Eng 11(5):301–310
29. Malinowsky B, Grønbæk J, Schwefel H-P, Ceccarelli A, Bondavalli A, Nett E (2012) Timed broadcast via off-the-shelf WLAN distributed coordination function for safety-critical systems. Proc. European Dependable Computing Conference (EDCC), pp 144–155, 8–11
30. Mills D (1992) RFC1305 - Network Time Protocol (Vers. 3) specification, implementation and analysis
31. Mock M, Nett E, Schemmer S (1999) Efficient reliable real-time group communication for wireless local area networks. In: Hlavicka J, Maehle E, Pataricza A (eds) Proceedings of the third European dependable computing conference on dependable computing (EDCC-3). Springer, Berlin, pp 380–400
32. NIST (1995) FIPS 180-1: Secure hash standard. US Department of Commerce, Washington DC
33. Office of Rail Regulation (2005) Annual report on railway safety. http://www.rail-reg.gov.uk/upload/pdf/296.pdf. Accessed 22 March 2012
34. Rivest R, Shamir A, Adelman L (1978) A method for obtaining digital signature and public key cryptosystems. Comm ACM 21:120–126
35. Römer K, Blum P, Meier L (2005) Time synchronization and calibration in wireless sensor networks. In: Stojmenovic I (ed) Handbook of sensor networks: algorithms and architectures. Wiley, New York, pp 199–237
36. Satyanarayanan M (2001) Pervasive computing: vision and challenges. Personal communications, vol 8, no 4. IEEE, New York, pp 10–17
37. Seminatore A, Ghelardoni L, Ceccarelli A, Falai L, Schultheis M, Malinowsky B (2012) ALARP (A railway automatic track warning system based on distributed personal mobile terminals). Transport Research Arena 2012 (TRA 2012) - Procedia Social and Behavioural Sciences, Elsevier, vol. 48, pp 2081–2090

38. Stallings W (2010) Cryptography and network security: principles and practice, 5th edn. Prentice Hall, Upper Saddle River
39. Thatte SM, Abraham JA (1980) Test generation for microprocessors. IEEE Trans Comput 29(6):429–441
40. Trimble R7 GNSS Base Station datasheet. http://www.trimble.com/trimbler7gnss_ds.asp. Accessed 20 August 2012
41. Trimble R8 GNSS receiver datasheet. http://www.trimble.com/trimbler8gnss_ds.asp. Accessed 22 March 2012
42. van de Goor AJ (1991) Testing semiconductor memories: theory and practice. Wiley, New York
43. Veitch D, Babu S, Pàsztor A (2004) Robust synchronization of software clocks across the Internet. In: Proceedings of the 4th ACM SIGCOMM conference on internet measurement. ACM, New York, NY, USA, pp 219–232
44. Verissimo P (2002) Travelling through wormholes: Meeting the grand challenge of distributed systems. In: Proceedings of the International Workshop on Future Directions in Distributed Computing. Bertinoro, Italy, pp 144–151
45. Verissimo P, Rodriguez L (2001) Distributed systems for system architects. Kluwer, Dordecht
46. Xen 4.0. http://wiki.xensource.com. Accessed 20 August 2012

Chapter 3
Self-Organizing Real-Time Services in Mobile Ad Hoc Networks

Yoshiaki Kakuda, Tomoyuki Ohta, and Miroslaw Malek

Abstract This chapter introduces a new concept of assurance networks and proposes a unified model and methodology for assurance networks, and their application to mobile ad hoc networks. First, we define assurance networks and present the four step design methodology, the assurance metric, and the unified design procedures. Next, we show a concrete methodology for real-time, self-organized control for autonomous clustering based on variation of mobility in mobile ad hoc networks. Finally, we suggest a number of application areas in which the proposed methodology of assurance networks can be used.

3.1 Assurance Networks

Along with rapid growth of the Internet cloud computing and ubiquitous networks, autonomous decentralized systems and cyber-physical embedded systems are connected to each other. Heterogeneous requirements are independently generated and the requirements themselves are frequently changing over the life of a system. Assurance in autonomous decentralized networks is defined as capability of guaranteeing functional and nonfunctional network properties such as dependability, security, timeliness and adaptivity even in the case of heterogeneous and changing requirements. Not surprisingly, in such environment assurance is a challenge. Techniques and methodologies supporting assurance including integration of various

Y. Kakuda (✉) • T. Ohta
Graduate School of Information Sciences, Hiroshima City University, 3-4-1 Ozuka-Higashi Asa-minami-ku, Hiroshima 731–3194, Japan
e-mail: kakuda@hiroshima-cu.ac.jp; ohta@hiroshima-cu.ac.jp

M. Malek
Institute for Informatics, Humboldt-Universität zu Berlin, Unter den Linden 6, D-10099 Berlin, Germany
e-mail: malek@informatik.hu-berlin.de

M.T. Higuera-Toledano et al. (eds.), *Self-Organization in Embedded Real-Time Systems*, DOI 10.1007/978-1-4614-1969-3_3,

technologies such as real time, fault tolerance, autonomy, mobility and intelligence will have to be incorporated in complex autonomous decentralized networks to make them trustworthy.

Since users of new generation networks demand and use various services and contents, the data volume in such networks is huge and frequently changing. In order to sustainably and stably respond to such user requirements and make service applications trustable through terminals and infrastructures, the new generation networks must guarantee assurance. This is a challenging research and design issue as technologies supporting assurance will have to be incorporated in the new generation networks.

The origins of assurance networks date back to responsive systems [2, 5, 8–14] whose aim was to guarantee timeliness even in presence of faults. In the parallel and distributed computing environments where computers are connected with each other through networks, abnormality of some computer may contaminate the entire network and cause much damage to other computers. In this situation, enhancement of performance for not only normal processing but also abnormal processing is required to mitigate the impact of faults leading to the essential concept of responsive systems.

The concept of responsive systems has been proposed by one of the authors in 1990 [8, 9] and responsive systems are defined as systems that integrate functions of fault-tolerant systems and real time systems in the parallel and distributed computing environment. In other words, the responsive systems are those that expected services can be timely executed even in the presence of faults. A unified design methodology for responsive systems was presented in [5, 8–10].

Assurance networks can be regarded as a generalization of responsive systems. Not only fault tolerance and real-time systems requirements are considered as in responsive systems but also security, autonomy, mobility, scalability and additional knowledge to support adaptivity are necessary for assurance networks [3]. Based on the definition of responsive systems, assurance networks are defined as those that provide timely services even when (1) the size of the networks changes (scalability issue), (2) the requirements and environments of the networks are dynamically changed (dynamicity and mobility issues), (3) cyber attacks occur (security issue) and (4) there are faults that may cause failures (fault tolerance issue) [4].

3.2 Assurance Networks Design

In order to design assurance networks, we propose a unified model and four-step design methodology as follows (see Fig. 3.1):

Step 1: Partition the overall network into disjoint subnetworks such that each subnetwork can be managed independently [15]. Please note that the traffic volume is key variable in a fixed network while the number of nodes is characterized in a mobile ad hoc network (MANET) and a peer to peer network.

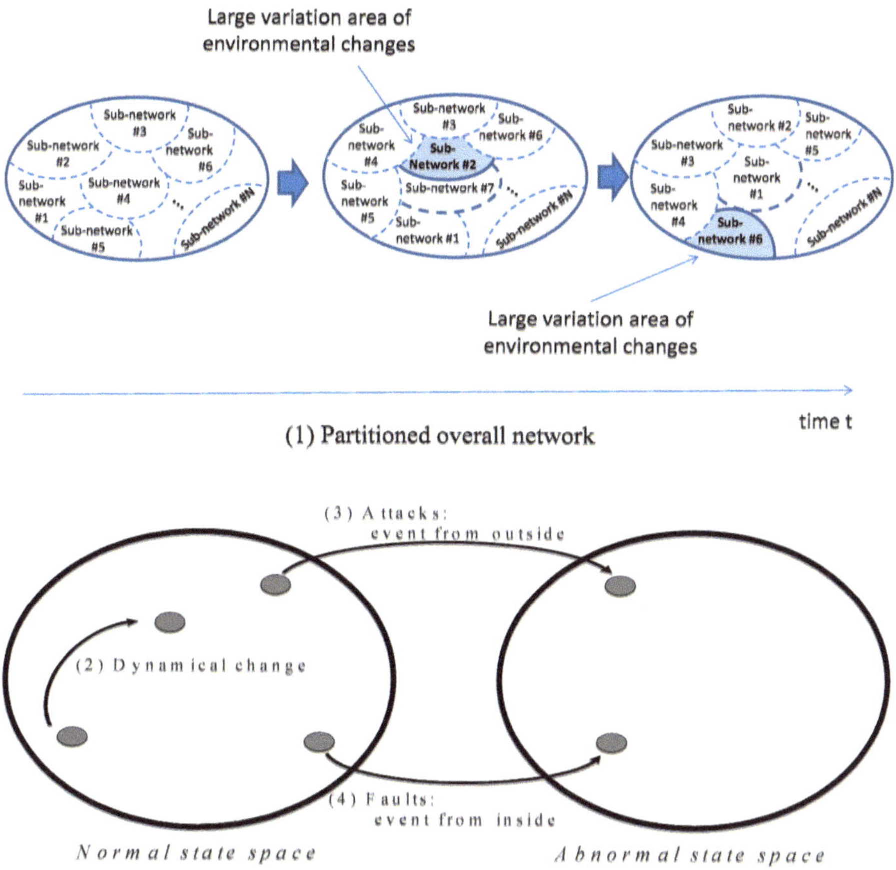

Fig. 3.1 Unified design model for assurance networks

The traffic volume and the number of nodes of each subnetwork are bounded by an upper bound U and a lower bound L where U is at least twice the size of L ($U \geqq 2L$). The traffic volume or the number of nodes of the subnetwork changes due to dynamical activities of nodes, attacks and failures of nodes in the subnetwork. Whenever the traffic volume or the number of nodes in a subnetwork becomes larger than U, the subnetwork is divided into two subnetworks each of whose size is bounded by U and L. Whenever the traffic volume or the number of nodes becomes smaller than L, the subnetwork tries to merge with one of its physically or logically neighboring subnetworks such that its size is larger than or equal to L and less than or equal to U. If it is not feasible the two neighboring subnetworks should merge and immediately split into two subnetworks to stay with the U and L bounds. This size control is necessary for effective management of each subnetwork. The two bounds U and L and the neighboring relation between subnetworks are determined depending on the given network properties.

Step 2: Model dynamical changes by transitions from normal states to the other normal states. Self-organized real-time control functions are incorporated into each subnetwork such that variation of the subnetwork requirements and/or environments is autonomously recognized by local information in the subnetwork, and control for the subnetwork is adaptively executed in real time according to the recognized pattern.

Step 3: Model attacks by transitions from normal states to abnormal states (they are caused by events from the outside of the networks and they are regarded as dynamical changes of the subnetwork environments except for their active abnormality). The attacks are recognized by local information in the subnetwork and recovery from abnormal states to normal states is performed by the real-time self-organized control functions in the subnetwork.

Step 4: Model faults by transitions from normal states to abnormal states caused by events from the inside of the networks and regard them as dynamical changes of the subnetwork environments except for their passive abnormality. The faults are recognized by local information in the subnetwork and recovery from abnormal states to normal states is performed by the real-time self-organized control functions in the subnetwork.

The assurance metric for an overall network is defined as resilience of the network performance to various network disturbance, environments and their variations. In order to provide timely services even when dynamical changes occur and faults possibly causing failure and attacks to security follow, assurance networks should be designed to withstand such situations. Since the overall network is partitioned into subnetworks in Step 1, the assurance metric for each subnetwork can be independently enhanced by real-time self-organized control functions incorporated into the subnetwork because dynamical change, faults possibly causing failure and attacks to security are uniformly modeled in Step 2, Step 3 and Step 4, respectively. Even when some subnetwork does not work due to its failure, one of its neighboring subnetworks can detect its failure and reconfigure it for recovery because all subnetworks partitioned in Step 1 are independent.

Figure 3.2 shows a conceptual diagram to represent assurance. In Fig. 3.2, overheads of controlling a network with low assurance do not vary with increase of the network complexity and performance of the network suddenly becomes worse when the network complexity increase due to lack of adaptability to environmental change. On the other hand, overheads of controlling a network with high complexity increase with increase of the network complexity and performance of the network becomes stable even when the network complexity varies because of adaptability to environmental change.

Based on the concept of assurance networks, we consider that networks with high assurance can keep high performance even when temporal and spatial environmental changes occur in the networks. Then, we define assurance metric (let it be denoted by A hereinafter) by integrating performance in terms of various environments, environmental changes, and spaces. Formula (3.1) for calculating assurance represents assurance metric for temporarily and spatially changing networks.

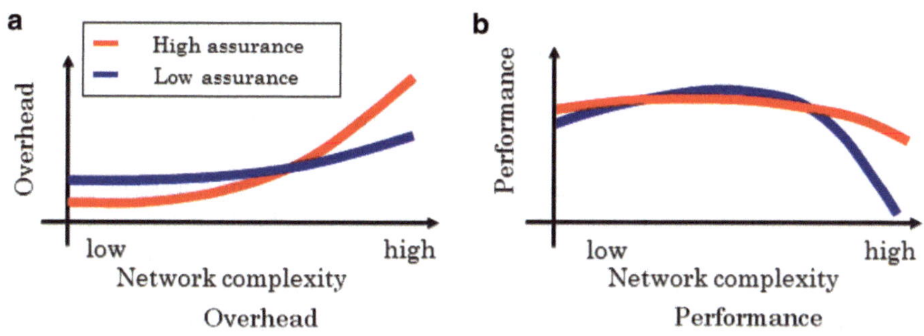

Fig. 3.2 Conceptual diagram for overhead and performance in high and low assurance networks with respect to complexity

$$A = \int_{y0}^{y1} \int_{x0}^{x1} \int_{v0}^{v1} \int_{v0}^{v1} (performance) dv \cdot d(w0 \rightarrow w1) \cdot dx \cdot dy \tag{3.1}$$

$\int_{v0}^{v1} dv$: Integration of performance for a diversity of network environments such as node mobility and density (let the environments be denoted by v). The larger this value is, the broader the network environments for which high performance is kept are.
$\int_{v0}^{v1} d(w0 \rightarrow w1)(v0 \leqq w0 \leqq v1, v0 \leqq w1 \leqq v1)$: Integration of performance for all changes (let them be denoted by $w0 \rightarrow w1$) of network environments. The larger this value is, the more high performance is maintained even when network environments change. Attacks to security and faults causing failures are regarded as dynamic changes of network environments. This element represents adaptability.
$\int_{y0}^{y1} \int_{x0}^{x1} dx \cdot dy$: Integration of performance for Euclidean spaces (let them be denoted by x, y). The larger this value is, the broader the spaces in which high performance is maintained are. Euclidean spaces can be substituted by logical spaces such as network topology. This element represents scalability.

In the above explanation, the formula for calculating assurance criteria is defined by integration of performance, however, if performance is given as discrete value, the formula can be defined by summation, maximum or minimum of performance. It is essential that assurance criteria consist of the above three elements. It is noticed that performance may include throughput, delay and some other metrics.

The unified design procedures for each subnetwork i are given in Fig. 3.3.

Nodes belonging to a subnetwork in a given time instance need not always remain to be members of the subnetwork forever. New nodes may join the subnetwork and some nodes may disappear. Accordingly, the number of nodes in the subnetwork can increase or decrease.

In order that all subnetworks do not overlap with each other and the size of all disjoint subnetworks becomes almost the same, a real-time self-organized control is applied to the structure of each subnetwork so that the size of each subnetwork is between the upper bound and the lower bound.

As shown in Fig. 3.3, if a subnetwork does not satisfy the criteria for partition, the following procedures are performed. If the size of the subnetwork becomes beyond the upper bound, the subnetwork is partitioned. If the size of the subnetwork becomes less than the lower bound, the subnetwork is merged with one of the neighboring subnetwork.

Since these partition and merger are done by exchanging messages with the subnetwork and between the subnetwork and its neighboring subnetwork, the overheads for exchanging messages are low and the processing time is short.

The above real-time self-organized procedures enable the structure of each subnetwork to be maintained in real-time so that the size of each subnetwork is almost the same even when the size of each subnetwork increases or decreases. As a result, since the load for controlling each subnetwork is distributed almost equally, the scalability of the overall network is assured.

As described in the above, if the overall network is partitioned into disjoint subnetworks and such a partitioned structure is dynamically maintained, real-time self-organized control that is independently applied to each subnetwork can be optimized and adaptable to environmental changes in each subnetwork even when they are different in each subnetwork.

Figure 3.3 also shows real-time self-organized control to three kinds of environmental changes in each subnetwork, that is, dynamical changes, attacks to security, and faults causing failure. First, detection techniques are executed to monitor the environmental changes. If no deviation to abnormal states is detected, the procedures are returned to maintenance of network partition. If it is detected, diagnosis methods for identifying their causes and impacts are executed. After the diagnosis, recovery methods are applied. If necessary, retry, reconfiguration, restart or some other methods are applied. As a result, if assurance and real-time properties are satisfied in each subnetwork, the procedures are returned to maintenance of network partition. Otherwise, detection, diagnosis and recovery techniques are applied to each subnetwork until assurance and real-time properties are satisfied. Since the detection, diagnosis and recovery technologies are real-time self-organized techniques, sustainable and stable services (assurance) can be guaranteed in real-time to each subnetwork.

The networks to which the unified design methodology by assurance technologies can be applied and be effective are non-reliable environments with many packet losses, temporarily and spatially mobile environments, environments whose resources are restricted, and environments in which real-time services are required. Such networks are not only wireless networks such as ad hoc networks and sensor networks, but also logical networks such as overlay networks and virtual networks. The difference between the methodologies that are applied to wireless networks and logical networks is physical and logical neighbor relation among physical and logical nodes. In case of virtual networks, using the methodology by assurance technologies virtual networks can be reorganized in real-time adaptively to change of network, computational and storage resources.

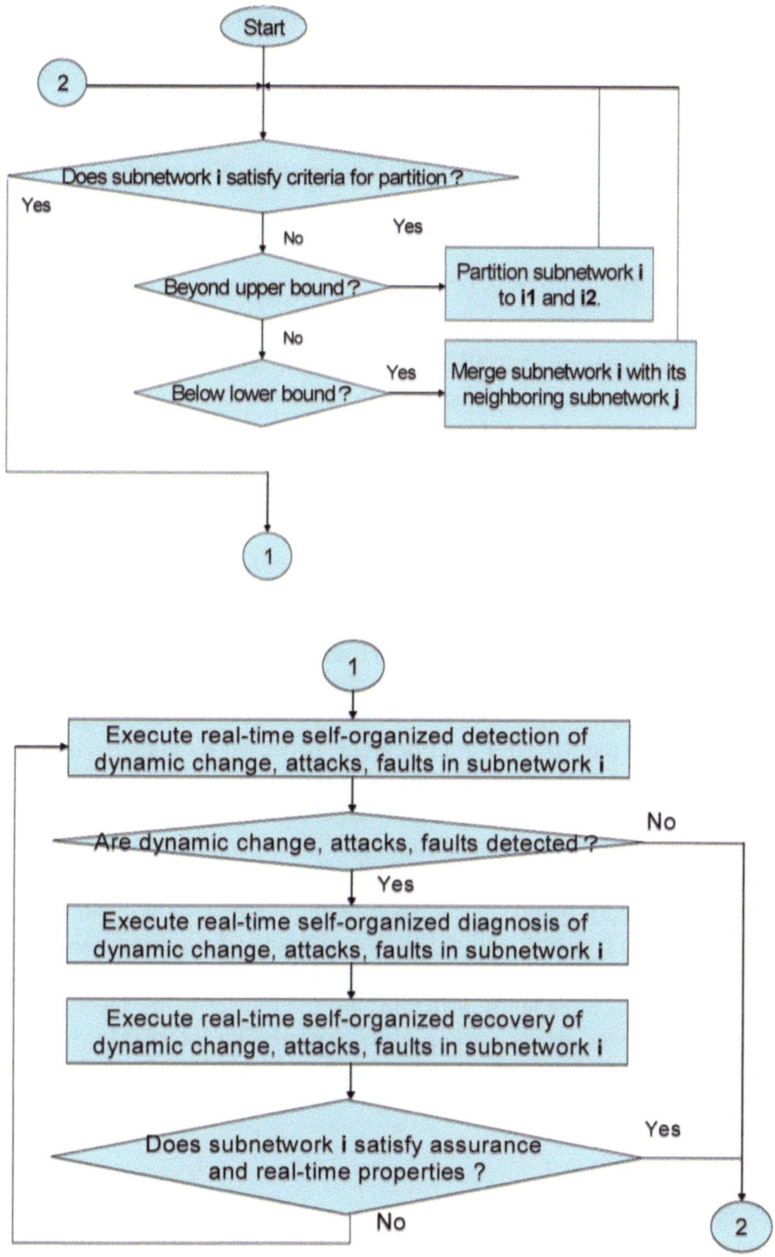

Fig. 3.3 Maintenance of network partition, detection, diagnosis, recovery of environmental changes

Once a model is developed as proposed the methodology shown in a case study on mobile ad hoc networks in Sects. 3.3 and 3.4 has to follow all the necessary steps to meet the requirements.

3.3 Case Study: Application of Assurance Networks to MANETs

3.3.1 MANETs

A mobile ad hoc network (MANET) is a network without any infrastructure, in which data are transmitted and received through wireless links between mobile terminals and forwarded by mobile terminals. Hereinafter, such mobile terminals are called nodes. Even when nodes move, data can be delivered from a source node to a destination node through a route of connected wireless links. In mobile ad hoc networks, it is generally difficult to predict speed and direction of node movement. Mobile ad hoc networks are one of the most extremely dynamic networks that vary their topology considerably with time.

In order to realize hierarchical routing for large-scale mobile ad hoc networks, we have developed autonomous clustering [17] with high scalability and adaptability. Autonomous clustering is referred to as the AMC (Adaptive Multi-hop Clustering) in a survey paper [21].

The variation of the MANET environments is caused by node movement, node battery consumption, radio transmission, packet transmission and other factors. As a result of the variation, node mobility, node density, packet transmission ratio and other parameters are continuously changing.

This section shows a methodology for real-time, self-organized control for autonomous clustering based on variation of mobility of each cluster. The more detailed description can be found in [7]. The mobility of a cluster is defined by an average node movement in the cluster. In the proposed control scheme, mobility of each cluster is independently estimated by the cluster using local information obtained in the cluster and according to it, transmission interval for broadcasting messages in the cluster is adaptively adjusted. As a result, assurance of autonomous clustering can be enhanced.

3.3.2 Autonomous Clustering

In autonomous clustering, all nodes in the overall network can be classified into disjoint sets of connected nodes called clusters. By treating a cluster as a virtual node, we can design a hierarchical routing, which consists of inter-cluster and intra-cluster routings. Nodes in each cluster are identified as follows: a clusterhead that manages the cluster, gateways that are connected to a neighboring cluster, and the remaining nodes called clustermembers. See Fig. 3.4.

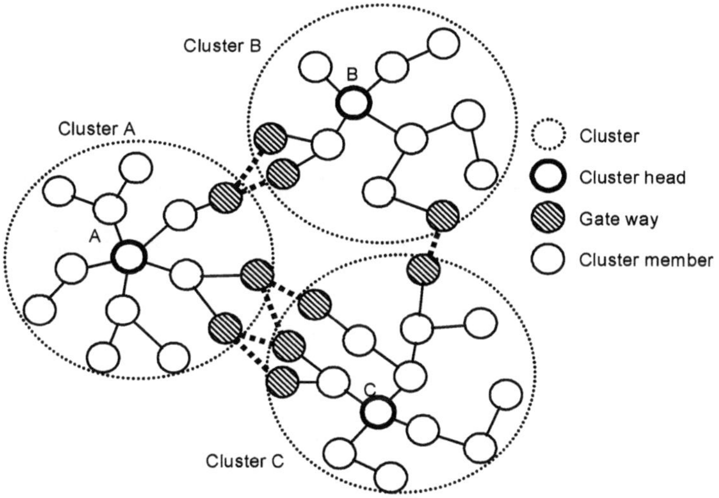

Fig. 3.4 Cluster organization in autonomous clustering

Each clusterhead constructs a spanning tree rooted at the clusterhead in the cluster by periodically broadcasting messages (called Member Packets (MEPs)) within the cluster and receiving reply messages (called Member Ack Packets (MAPs)). Through gateways, messages are delivered to neighboring clusters. Each clusterhead can collect information on the cluster and its neighboring clusters by MEPs and MAPs.

In autonomous clustering, the identification (ID) of a cluster to which a node (let it be v) belongs is determined by the IDs of nodes which are neighbors of the node v. Using this control scheme, each cluster can be kept even when nodes move around the network though the cluster to which some nodes belong in the cluster may be changed. See Fig. 3.5. This is a self-organized control of autonomous clustering as a node level, in which the cluster structure can be maintained even when some nodes in the cluster are changed. Adaptability of autonomous clustering is obtained by this fundamental self-organized control.

The size of a cluster can increase or decrease because autonomous clustering has the node level self-organized control. In order to keep the size of each cluster between the upper and lower bounds given beforehand, clusters are divided if the cluster size is over the upper bound and merged if it falls below the lower bound. This is another self-organized control of autonomous clustering at a cluster level. Since, even when the network becomes large-scale, each clusterhead has only to manage the limited size of the cluster, autonomous clustering assures scalability of mobile ad hoc networks.

In autonomous clustering, a node that manages a cluster periodically broadcasts MEPs within the cluster and receives MAPs. In the received messages, the number and IDs of all nodes and the numbers of nodes in each neighboring cluster are

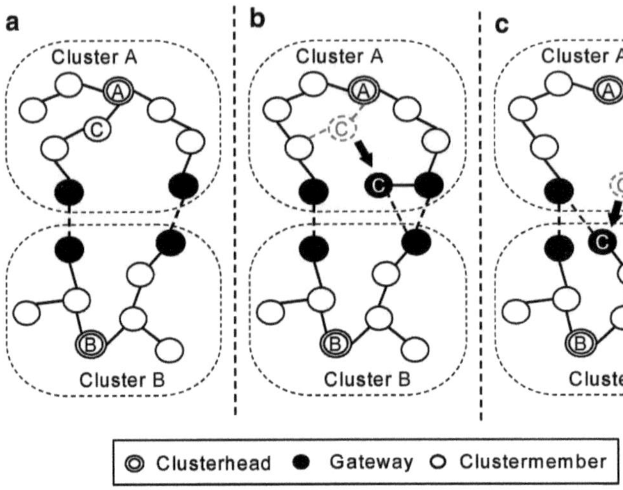

Fig. 3.5 Node level self-organized control

included. Based on this information, clusters are divided or merged so that the size of each cluster is kept between the given upper and lower bounds.

3.3.3 Hierarchical Routing Hi-AODV

In the autonomous clustering-based hierarchical routing, a route between each pair of a source node and a destination node is divided into subroutes between clusterheads via gateways in the neighboring clusters, a subroute between a clusterhead and a source node in the cluster and a subroute between a clusterhead and a destination node in the cluster. Since these subroutes are parts of the spanning trees rooted at the clusterheads, they can be independently constructed in each cluster. In autonomous clustering, each clusterhead can change the spanning tree adaptively to variation of the cluster topology due to node movement. The subroutes can be maintained in each cluster.

Hi-AODV consists of an intra-cluster routing for each cluster and an inter-cluster routing among clusters. The intra-cluster routing is based on the spanning tree rooted at a clusterhead. Data packets are transferred from a source node to the clusterhead through a subroute on a spanning tree and forwarded toward a destination node through another clusterheads. Finally, the data packets are transferred from a clusterhead to the destination node through a subroute on a spanning tree. The inter-cluster routing is based on AODV by regarding clusters as a virtual node. Data packets between clusterheads in the neighboring clusters are transferred through gateways to the neighboring clusters.

More detail description of unicast and hierarchical routings can be found in [19] and [18], respectively.

3.3.4 Real-Time Self-Organization

In autonomous clustering, MEPs are broadcast from a clusterhead and in the process of receiving MAPs a spanning tree rooted at the clusterhead is constructed. Data packets are transferred through routes on the spanning tree rooted at the clusterhead. In autonomous clustering, MEPs are broadcast periodically with a constant transmission power per packet to construct a spanning tree.

However, if any node in the cluster does not move, the clusterhead needs not to reconstruct the spanning tree rooted at the clusterhead. On the other hand, if some nodes in the cluster move fast, the clusterhead should frequently reconstruct the spanning tree by sending MEPs in a short interval.

In this section, we show a methodology for real-time self-organized control to change the interval for transmission of packets adaptively to variation of mobility of each cluster.

3.3.5 Real-Time Self-Organized Control Scheme Based on Cluster Mobility

Since nodes move around mobile ad hoc networks, a spanning tree rooted at a clusterhead does not always work for transmission of packets. If transmission interval of MEPs becomes long, the spanning tree cannot be constructed in time to compensate for node movement. As a result, data packets cannot be delivered along a route on the spanning tree and the delivered data packet ratio goes down. On the other hand, if transmission interval of MEPs becomes short, MEPs are broadcast even when reconstruction of the spanning tree is not necessary. As a result, many redundant MEPs and MAPs are transmitted and network resource is consumed in vain. Therefore, the transmission interval of MEPs needs to be adjusted independently in each cluster depending on mobility of the cluster.

We denote the transmission interval as a period from t_j to $t_{j+1} (t_j < t_{j+1})$ and we assume that clusters are not divided and merged during that period. The cluster mobility $D(v_i)$ for each node v_i in a cluster can be classified into the intracluster mobility and the intercluster mobility for node v_i in the cluster. The intracluster mobility $Dintra(v_i)$ for each node v_i in a cluster is defined as the number of nodes which open or close a neighboring relation to node v_i between t_j and t_{j+1}. The intercluster mobility $Dinter(v_i)$ for each node v_i in a cluster is defined as the number of nodes which enter into or exit from the cluster between t_j and t_{j+1} while keeping a neighboring relation to node v_i during that period. $D(v_i)$, $Dintra(v_i)$, and $Dinter(v_i)$ are described in the following formulas.

$$D(v_i) = w1 \times Dintra(v_i) + w2 \times Dinter(v_i)$$
$$Dintra(v_i) = |(N(v_i,t_j) \cup N(v_i,t_{j+1})) - (N(v_i,t_j) \cap N(v_i,t_{j+1}))|$$
$$Dinter(v_i) = |(N(v_i,t_j) \cap N(v_i,t_{j+1})) \cap ((C(v_i,t_j) \cup C(v_i,t_{j+1})) - (C(v_i,t_j) \cap C(v_i,t_{j+1})))|$$

Here, $w1$ and $w2$ are weights to indicate significance for $Dintra(v_i)$ and $Dinter(v_i)$, respectively ($w1 + w2 = 1$). $N(v_i,t_j)$ denotes a set of nodes which have a neighboring relation to node v_i at time t_j. $C(v_i,t_j)$ denotes a set of nodes which have a neighboring relation to node v_i and which belong to the same cluster as that of node v_i at time t_j.

Based on the neighboring node information, each node calculates $D(v_i)$ and delivers it to the clusterhead through the spanning tree. Then, the clusterhead calculates the maximum of $D(v_i)$ divided by the transmission interval for all nodes in the cluster and estimates it as the cluster mobility. The estimated cluster mobility is denoted by *Dmax*, which is described in the following formula.

$$Rmax = max\ D(v_i)/(t_{j+1} - t_j) \text{ for all nodes } v_i$$

We will show examples for $Dintra(v_i)$ and $Dinter(v_i)$ using Fig. 3.6. In the figure, a circle denotes a node, a solid line that connected two circles denotes their neighboring relation, a dashed line denotes a boundary of a cluster, and an arrow denotes node movement. In Fig. 3.6a, nodes B and D have neighboring relations to node A at time t_j, however, they move and, as a result, do not have any neighboring relations to node A at time t_{j+1}. Node H does not have any neighboring relation to node A at time t_j, however, after it moves, it has a neighboring relation to node A at time t_{j+1}. Node A thus has three changes of neighboring relations between times t_j and t_{j+1}. Then, Dintra(A)=3. The cluster to which node C belongs changes, however, node C keeps a neighboring relation to node A between times t_j and t_{j+1}. ï£¡@This change does not affect Dintra(A). In Fig. 3.6b, node A keeps neighboring relations to nodes C and F between times t_j and t_{j+1}, however, nodes C and F move and, as a result, the clusters to which they belong change. Then, Dinter(A)=2. Node B moves and, as a result, the cluster to which node B belongs changes between times t_j and t_{j+1}. However, node B also closes the neighboring relation to node A during that period. This change does not affect Dinter(A).

$N(v_i,t_j)$ and $C(v_i,t_j)$ are local information, which can be obtained by sending MEPs and receiving MAPs within the cluster. If the interval for transmitting MEPs is adjusted for variation of this local information, a real-time self-organized control can be realized.

Let $I_0, I_1, \ldots, I_m$ be $m+1$ transmission intervals. Here, $I_0 > I_1 > \ldots > I_m$. Let us define the m mobility thresholds that correspond to transmission intervals follow $R_0, R_1, \ldots, R_{m-1}$ where $R_0 < R_1 < \ldots < R_{m-1}$.

Suppose that the current transmission interval is $I_i (i < m)$ and $T_a < Rmax < T_{a+1}$. If, for the following n consecutive intervals where n is a positive integer, $Rmax >$

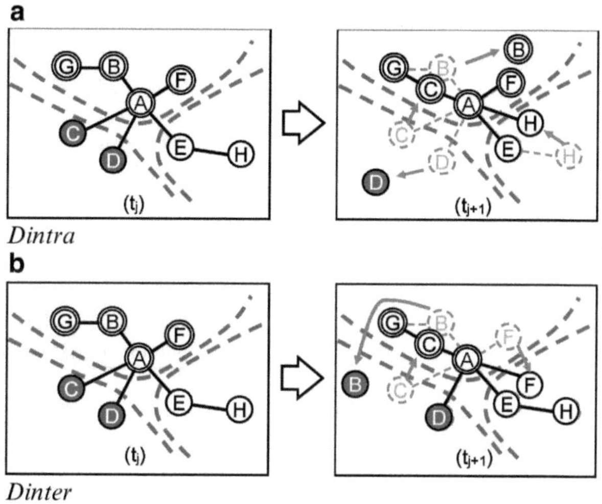

Fig. 3.6 Examples for *Dintra* and *Dinter*

T_{a+1} is kept, decrease the transmission interval and let the transmission interval be I_{i+k} where $i+k$ is less than or equal to m.

Suppose that the current transmission interval is $I_j(j>0)$ and $T_b < Rmax < T_{b+1}$. If, for the following n consecutive intervals where n is a positive integer, $Rmax < T_b$ is kept, increase the transmission interval and let the transmission interval be I_{j-r} where $j-r$ is greater than or equal to 0.

3.4 Evaluation of Real-Time Self-Organized MANETs

We evaluate the performance of Hi-AODV based on autonomous clustering scheme (shortly, ac scheme) and Hi-AODV based on autonomous clustering with real-time self-organized control scheme (shortly, ac-rts scheme).

3.4.1 Simulation Environment

The simulation runs for a large network where the number of nodes is 150. Tables 3.1 and 3.2 show common parameters and parameters for the ac-rts and ac schemes, respectively. In Table 3.1, SD pair means a pair of the source node and the destination node which communicate with each other. Source nodes start sending data packets from 30 s on every 0.25 s. This is because it takes about between 10 s and 20 s to make autonomous clustering stable.

In this simulation evaluation, we performed two types of simulation experiments to evaluate the ac-rts scheme on variation of the cluster mobility. The first type (Type 1) is mobility fixed with time and the second type (Type 2) is mobility varied with time.

Table 3.1 Common parameters for simulation environment

Simulator	QualNet ver. 3.9.5
Size of field	2000×1300 (m^2)
Mobility model	Random waypoint model
Pause time	0(s)
Simulation time	330 (s)
Radio propagation model	Two-ray ground
Bandwidth	54 (Mbps)
MAC protocol	IEEE 802.11
Packet size	512 (byte)
Interval of sending data packets	0.25 (s)
Number of SD pairs	30

Table 3.2 Parameters used for ac and ac-rts schemes

Cluster size (L, U)	(10, 50)
Interval of sending a MEP	1.0 (s)–3.0(s)
Weights $(w1, w2)$	(0.5, 0.5)

3.4.2 Simulation Experiment I (Type 1)

We conducted simulation experiments in case of the fixed node mobility. In the simulation environment, the field size is 2,000 m by 1,300 m and the number of nodes is 150. The node speeds are 1 m/s, 5 m/s, 10 m/s. The random waypoint model [1] is adopted as the mobility model. In this model, at every instant, a node randomly chooses a destination and moves towards it with a velocity chosen uniformly randomly from $[0, Vmax]$, where $Vmax$ is the maximum allowable velocity for every mobile node. The au-rts scheme is one which simplifies the control scheme provided in Sect. 3.5 as follows.

Suppose that the current transmission interval is $I_i (i < m)$ and $R_a < Rmax < R_{a+1}$. If, for the following interval, $Rmax > R_{a+1}$, decrease the transmission interval and let the transmission interval be I_{i+1}.

Suppose that the current transmission interval is $I_j (j > 0)$ and $R_b < Rmax < R_{b+1}$. If, for the following interval, $Rmax < R_b$, increase the transmission interval and let the transmission interval be I_{j-1}.

We set the thresholds and the transmission intervals as follows. $R_0 = 0.375$, $R_1 = 0.75$, $R_2 = 1.5$, $R_3 = 3.0$ and $I_0 = 3.0$, $I_1 = 2.5$, $I_2 = 2.0$, $I_3 = 1.5$, $I_4 = 1.0$. Since R_i's $(i = 0, 1, 2, 3)$ are thresholds on ratio of mobility change, we set $R_j / R_{j+1} = 2 (j = 0, 1, 2)$. Since I_i's $(i = 0, 1, 2, 3, 4)$ are transmission intervals, we set $I_{j+1} - I_j = 0.5 (j = 0, 1, 2, 3)$. Base on the preliminary simulation experiments, these parameters are determined. The detail reason why they are determined is shown in [7].

Figures 3.7 and 3.8 represent the ratio of delivered data packets and the total size of control packets, respectively, for a combination of the average node speeds 1 m/s, 5 m/s, 10 m/s and the ac-rts scheme and the ac schemes in which the interval of sending MEPs is 1, 1.5, 2, 2.5, 3 s. Here, the control packets are those for

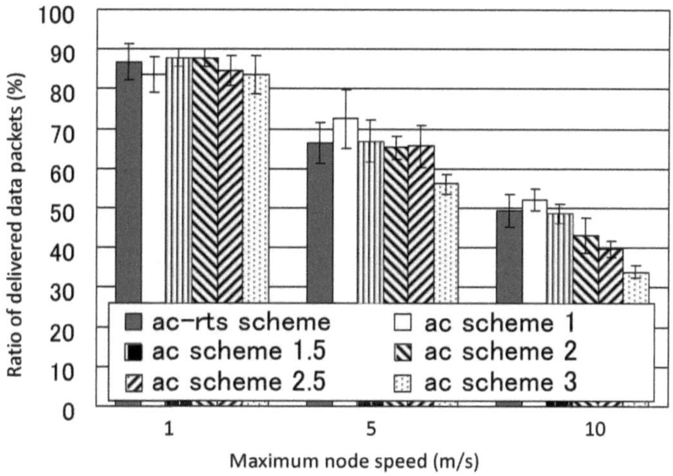

Fig. 3.7 Data packet delivery ratio

intracluster routing such as MEP and MAP. In these figures, the ac scheme with $k(k = 1, 1.5, 2, 2.5, 3)$ denotes one in which the interval of sending MEPs is 1, 1.5, 2, 2.5, 3 s, respectively, and each error bar denotes 90 % confidence interval.

First, we evaluate the ac schemes with different transmission intervals. Figure 3.7 shows that along with increase of transmission intervals, the ratio of delivered data packets decreases. As the maximum data speeds become high, the decreasing ratios become large. This result indicates that each cluster should control the transmission interval short enough to timely construct the clusterhead-based spanning tree for network environments of high node mobility. Figure 3.8 shows that along with increase of transmission intervals, the total size of control packets decreases regardless of the maximum node speeds. The node mobility does not much affect the total size of control packets. This result indicates that the transmission interval should be as large as possible from the viewpoint of control packets overhead.

Next, we evaluate the ac-rts scheme by comparing it with the ac schemes. Figure 3.7 shows that the ratio of delivered data packets of the ac-rts scheme is slightly smaller than that of the ac scheme with transmission interval 1 s, however, compatible with the ac scheme with transmission interval 1.5 s. Figure 3.8 shows that, as the maximum node speed becomes high, the total size of control packets in the ac-rts scheme increases. This is the reason why in the ac-rts scheme each cluster can shorten the transmission interval to construct the clusterhead-based spanning tree on time for network environments of high node mobility. The total size of control packets in the sc-rts scheme is much smaller than that in the ac scheme with transmission interval 1 s and compatible with that in the ac scheme with transmission interval 2 s even when the maximum node speed is 10 s.

Therefore, although the ac scheme with the shortest transmission interval has the best throughput with the largest overhead, the ac-rts scheme can maintain as good

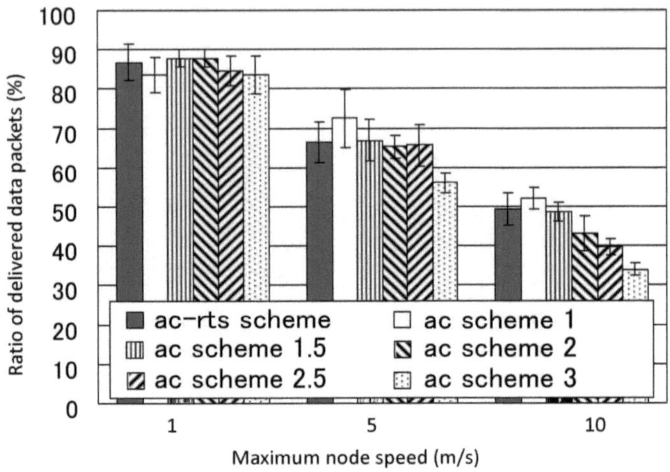

Fig. 3.8 Control packets size

throughput as the ac scheme regardless of the node mobility while keeping low overhead even in case of high node mobility. Although throughput and overhead of the proposed scheme are closest to the ac scheme with $k = 1.5$, the ac-rts scheme has an attractive capability that it is resilient to variation of node mobility. This will also be confirmed in the next subsection.

3.4.3 Simulation Experiment II (Type 2)

In the following simulation experiments, the node mobility changes along with the elapsed time. We investigate the following two cases. In the Case 1, from the beginning all nodes move at speed 1m/sec, from 50 s to 200 s, all nodes move at speed 10 m/s, from 200 s to 300 s, all nodes move at speed 1m/sec. In the Case 2, from the beginning all nodes move at speed 1m/sec, from 60 s to 120 s, all nodes move at speed 5 m/s, from 120 sec to 180 s, all nodes move at speed 10 m/s, from 180 s to 240 s, all nodes move at speed 5 m/s, from 240 s to 300 s, all nodes move at speed 1 m/s.

Figure 3.9 represents the number of MEPs along with the elapsed simulation time for Case 1 and Case 2. In these figures, the horizontal axis denotes the simulation time and the vertical axis denotes the average number of MEPs per s for every 10 s. These figures confirm that the ac-rts scheme adaptively adjusts the transmission interval to change of the node speeds. Figure 3.10 represents the number of delivered data packets per 2 s (we call it throughput) for the ac-rts scheme and the ac schemes with transmission intervals 1 s, 2 s, 3 s. From the beginning to about 30 s, the throughput increases in both schemes because many control packets are transmitted for constructing stable clusters in the network. Except for this period, the varied

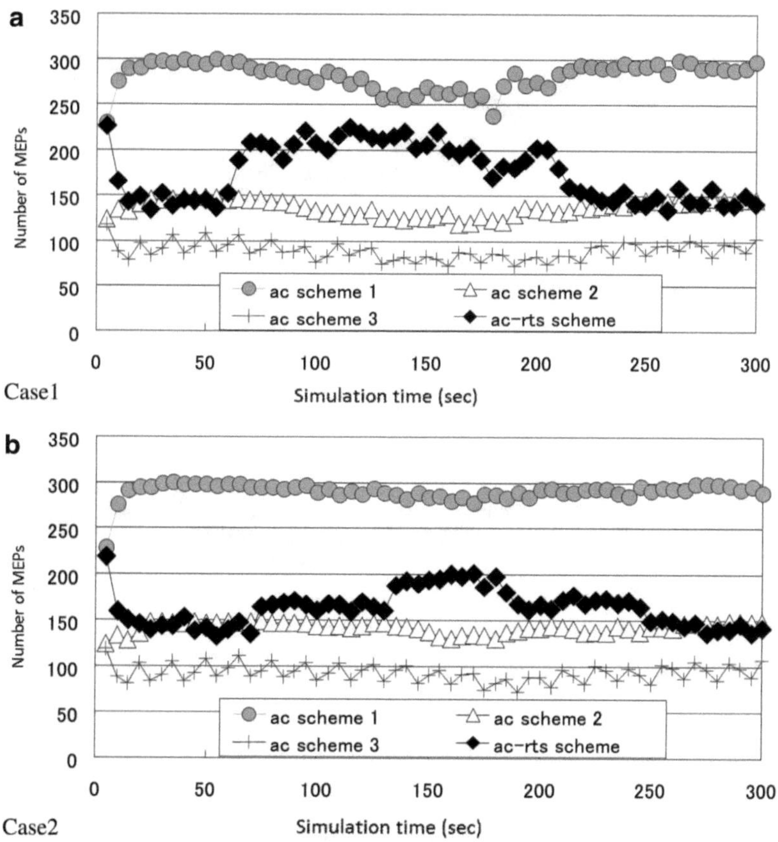

Fig. 3.9 Number of MEPs with elapsed simulation time

throughput of the ac-rts scheme is almost same as that of the ac scheme with transmission interval 1 s and is higher than that of the ac scheme with transmission intervals 2 s and 3 s. The throughput decreases whenever the node speed is high and the throughput increases whenever the node speed is low. In these cases, the throughput decreasing and increasing ratios of the ac-rts scheme is almost same as those of the ac scheme with transmission interval 1. On the other hand, the throughput decreasing and increasing ratios of the ac schemes with transmission intervals 2 s and 3 s are small. This result is clear especially from 200 s on in the Case 1. Based on the above observation, we can conclude that, even when the node mobility varies with time, the ac-rts scheme keeps high throughput with low overhead.

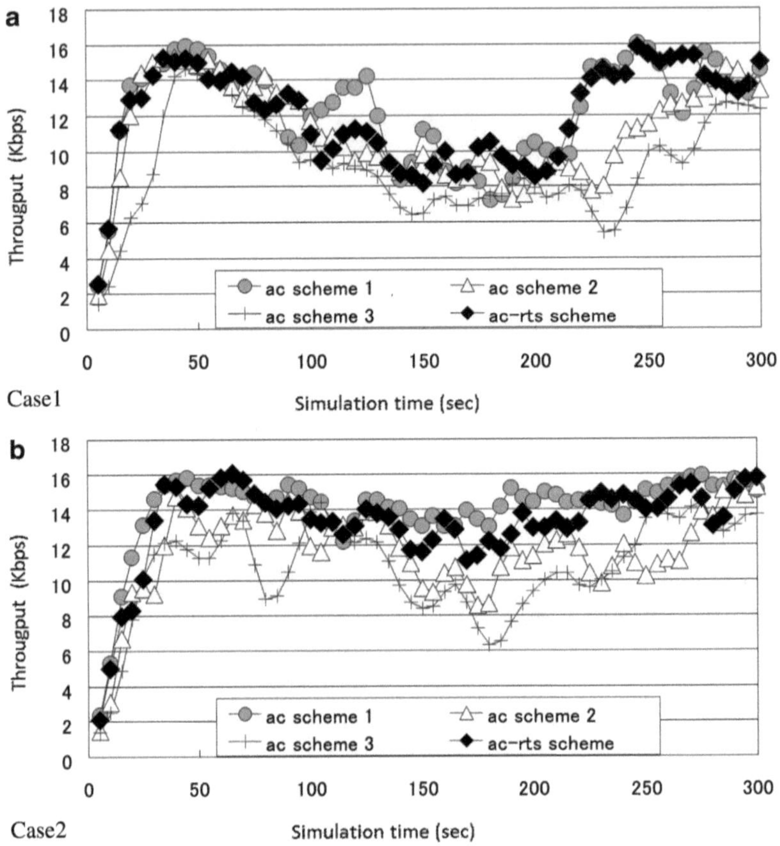

Fig. 3.10 Throughput with elapsed simulation time

3.5 Concluding Remarks

This chapter has introduced a new concept of assurance networks and proposed a unified design model and methodology for assurance networks and its application to mobile ad hoc networks. The concept and the model can be applied to many services and applications requiring high dependability, security, timeliness and adaptivity. As a case study, a methodology for real-time, self-organizing autonomous clustering for mobile ad hoc networks has been presented. In the proposed methodology, each clusterhead adjusts the interval of broadcasting a message within each cluster according to mobility of the cluster. The simulation experimental results showed that the presented autonomous clustering is resilient to a diversity of node mobility and its variation. We have also applied the proposed methodology to enhance assurance of safe and reliable data transfer [6] in wireless sensor networks and assurance of fault-tolerant routing [20] in MANETs. We believe that the proposed unified model and an exemplary applications of the proposed methodology to mobile ad hoc networks are the initial important steps in responding to the assurance challenge in new generation networks.

Acknowledgements This research was supported by the Ministry of Education, Science, Sports and Culture of Japan under Grant-in-Aid for Scientific Research (B) (No.21300028), the National Institute of Information and Communications Technology, Japan under Early-concept Grants for Exploratory Research on New-generation Network (No.145-9) and Hiroshima City University.

References

1. Breslau L, Estrin D, Fall K, Floyd S, Heidemann J, Helmy A, Huang P, McCanne S, Varadhan K, Xu Y, Yu H (2000) Advances in network simulation. IEEE Comput 33(5):59–67
2. Fussell DS, Malek M (eds) (1995) Responsive computer systems: steps towards fault-tolerant real-time systems, ISBN: 0792395638. Kluwer, Dordecht
3. Kakuda Y (2010) Assurance networks: concepts, technologies, and case studies, keynote speech, the second international symposium on multidisciplinary emerging systems and networks (MENS2011). In: Proceedings of the 2010 symposia and workshops on ubiquitous, autonomic and trusted computing, IEEE Computer Society, Oct 2010, pp 311–315
4. Kakuda Y, Malek M (2011) A unified design model for assurance networks and its application to mobile ad hoc networks. In: Proceedings of the 10th international symposium on autonomous decentralized systems (ISADS2011). 10th international workshop on assurance in distributed systems and networks (ADSN2011), IEEE Computer Society, pp 637–646
5. Kakuda Y, Kikuno T, Malek M, Saito H (1992) A unified approach to design of responsive protocols. In: Proceedings of the 1992 IEEE workshop on fault-tolerant parallel and distributed systems, Amherst, MA, 6–7 July 1992, pp 8–11
6. Kohno E, Okazaki T, Takeuchi M, Ohta T, Kakuda Y, Aida M (2012) Improvement of assurance including security for wireless sensor networks using dispersed data transmission. J Comput Syst Sci, Elsevier, vol.78, no.6, November 2012, pp 1703–1715. doi:10.1016/j.jcss.2011.10.018. (http://dx.doi.org/10.1016/j.jcss.2011.10.018)
7. Kakuda Y, Ohta T, Oda R (2011) A methodology for real-time self-organized autonomous clustering in mobile ad hoc networks. Concurrency and computation: practice and experience. Wiley, New York. doi:10.1002/cpe.1792
8. Malek M (1990) Responsive systems – a challenge for the nineties, keynote address. In: Proceedings of Euromicro '90, sixteenth symposium on microprocessing and microprogramming. Microprocessing and Microprogramming, vol 30, August 1990. Amsterdam, The Netherlands, North Holland, pp 9–16
9. Malek M (1991) Responsive systems: a marriage between real time and fault tolerance, keynote address. In: Proceedings of the fifth international conference on fault-tolerant computing systems. Informatik-Fachberichte, vol 283, September 1991. Springer, Nuernberg, Germany, pp 1–17
10. Malek M (1993) A consensus-based model for responsive computing. In: IEICE transactions on information and systems. Special issue on responsive computer systems, vol E76-D, no 11, November 1993, IEICE, pp 1319–1324
11. Malek M (ed) (1994) Responsive computing, special issue of real time systems. Int J Time-Critical Comput Syst 7(3), ISBN: 0792395115
12. Malek M (1994) A consensus-based framework and model for the design of responsive computing systems. In: Koob G, Lau c (eds) Ultradependable parallel and distributed computing. Kluwer, Dordecht

13. Malek M (1995) Omniscience, consensus, autonomy: three tempting roads to responsiveness. In: Proceedings of The 14th symposium on reliable distributed systems, keynote address, Bad Neuenahr, September 1995
14. Malek M, Polze A, Werner M (1996) The unstoppable orchestra – a responsive distributed application. In: Proceedings of the third international conference on configurable distributed systems, May 1996. IEEE Computer Society, Annapolis, pp 154–160(7), ISBN 0-8186-7395-8
15. Mori K (2008) Autonomous decentralized system and its strategic approach for research and development, invited paper, special issue on autonomous decentralized systems theories and application deployments. In: IEICE transactions on information and systems, vol E-91-D, no 9, Sept 2008, pp 2227–2232
16. NICT: Diversity & Inclusion (2009) Networking the future, technology strategies for a new-generation network, Interim Report, http://nwgn.nict.go.jp/report/NWGN-RD-Strategy-NICT-Report-V1-2009-EN.pdf. Accessed March 2009
17. Ohta T, Inoue S, Kakuda Y (2003) An adaptive multihop clustering scheme for highly mobile ad hoc networks. In: Proceedings of the 6th IEEE international symposium on autonomous decentralized systems (ISADS2003), pp 293–300
18. Ohta T, Fujimoto M, Oda R, Kakuda Y (2004) A class of hierarchical routing protocols based on autonomous clustering for large mobile ad hoc networks. IEICE Trans Comm E87-B(9):2500–2510
19. Perkins C, Belding-Royer E, Das S (2003) Ad hoc on-demand distance vector (AODV) routing. In: IETF RFC 3561. http://www.ietf.org/
20. Takeuchi M, Kohno E, Ohta T, Kakuda Y (2016) Improving assurance of a sustainable route-split MANET routing through adaptable node battery exhaustion. Telecomm Syst J, Springer, vol.61, no.4, 2016
21. Yu JY, Chong PHJ (2005) A survey of clustering schemes for mobile ad hoc networks. IEEE Comm Surv Tutor 7(1):32–48

Chapter 4
Swarm Robotic Time Synchronization for Object Tracking

Yara Khaluf, Emi Mathews, and Franz Josef Rammig

Abstract Wireless sensorand ad-hoc networks have been integrated into many self-organized tasks, including self-organized real-time tasks. Swarm robotics is a new field of research, which offers a set of advantages like motion, redundancy, flexibility, etc. compared to both sensor networks and ad-hoc ones. On the other hand, there are some difficulties in directly using swarm robotics for these kinds of tasks without modifying or even extending some of the strategies and protocols used in wireless sensor and ad-hoc networks. Time synchronization may serve as a prominent example of extensions needed to fit swarm systems. Our article focuses on employing swarm robotics in self-organized object tracking tasks. We develop a new strategy for overcoming the effect of the high degree of motion in swarm robotics via applying time synchronization protocols.

4.1 Introduction

Swarm robotics may be seen as an extension of wireless sensor networks (WSN). Like wireless sensor nodes, swarm robots are capable of performing various sensing or surveillance missions. In contrast to sensor nodes which usually are deployed statically, robots of a swarm are capable to move around, either as part of the deployment or after the deployment. This offers additional freedom, allowing more efficient mission accomplishment. Like sensor networks, swarm robotics constitute a large number of relatively simple nodes, each of them having limited capabilities. In the case of swarm robots these capabilities include the ability to move around, however with a rather limited speed due to limited battery capacity

Y. Khaluf (✉) • E. Mathews • F.J. Rammig
Heinz Nixdorf Institute, University of Paderborn, Fürstenallee 11, 33102 Paderborn, Germany
e-mail: yara@hni.uni-paderborn.de; emi@hni.uni-paderborn.de; franz@uni-paderborn.de

M.T. Higuera-Toledano et al. (eds.), *Self-Organization in Embedded Real-Time Systems*, DOI 10.1007/978-1-4614-1969-3_4,

and limited range. The nodes of a robot swarm can sense within a certain range and communicate within a different, usually larger range than the sensing range to their neighboring nodes. Based on this communication there is the potential of a complex macro-level behavior to emerge from a collection of simple micro-level behaviors.

Taking inspiration from nature, especially from the behavior of social insects, like ant colonies, is relatively widespread in the area of swarm robotics. Such animals show a highly complex macro-level behavior based on a certain coordination of simple micro-level behaviors. Other animals may also show cooperative behavior resulting in highly sophisticated macro-level capabilities. Cooperative hunting may serve as an example.

Whenever there is cooperation, an adequate degree of coordination has to be present. The individual objects involved have to share certain resources, including space and time-lines. This implies restrictions concerning local behaviors, may it be in movement or manipulation or even in access to communication channels. Therefore a certain degree of coordination is needed to avoid deadlocks, task failures, or even severe damages.

Although the basic concepts of swarm robotics have been developed for swarms of very simple robots, basic concepts can be transferred to multi-robot systems. These are collections of more complex robots. Such multi-robot systems could be used in various application areas like exploration, surveillance, or rescue operations and achieve improved performance and reliability in comparison to single-robot solutions. Multi-robot systems show preferable properties concerning scalability, flexibility, robustness, and fault tolerance due to the avoidance of a single point of failure.

The scenario we are coping with is that of one or more robotic swarms tracking moving targets in a complex and dynamically changing environment. In addition to tracking the movement of targets, the swarms may collect additional information about the objects and the environment. Such scenarios are rather common in military, security, and other applications. There is literature available about the scenario described above, e.g. [9, 10, 19]. In these papers multi-robot systems are used. An alternative may be to deploy swarms of simple robots. This approach, however, is challenging due to the limited capabilities of the individual robots used. Their restricted speed, sensing, and communication range impose problems to be solved. On the other hand the increased number of individual robots together with the ability to communicate opens novel solution options.

Applications such as localizing a rapidly moving target by cooperative triangulation or cooperatively handling some delicate materials, clearly indicate the necessity of time synchronization. We are assuming rather simple individual robots which even have no access to GPS or comparable means as a common time source. As we are considering more than one swarm performing the mission, time synchronization among the swarms on top of intra-swarm synchronization is an additional challenge. Moreover, time synchronization has to be carried out not only at the beginning of the mission but also rather frequently during the execution of the task. This is necessary to make sure that events are registered chronologically correct among all robots involved in a certain task.

In this chapter we are concentrating on this problem of time synchronization. Other important aspects like the Multi-Robot Task Allocation problem (see [3]) are not addressed. Concerning optimizing area coverage and target tracking just the state-of-the-art is shortly mentioned. Our own research in this area is described in [11].

The remainder of this chapter is organized as follows: Sect. 4.2 provides a description of the new time synchronization technique for sets of robot swarms. Section 4.3 introduces a case study scenario where the proposed time synchronization technique could be applied. In Sect. 4.4, we present experimental results. Related approaches known from the literature are then discussed in Sect. 4.5. Finally, Sect. 4.6 summarizes our solutions for the object tracking problem.

This chapter is partly an extension of a contribution published in the proceedings of SORT 2011[11]. Section 4.2, however, is completely new with respect to [11].

4.2 Time Synchronization in Swarm Robotics: New Strategy

Time synchronization being a main concept in wireless networks for many application domains, it will be a main concept in swarm robotics systems as well. A large set of well-known time synchronization protocols were developed for the wireless communication specifically for wireless sensor networks see Sect. 4.5. All of these protocols can also be used for time synchronization in swarm robotics. However, swarm robotics systems have some different properties in comparison to wireless sensor networks. One of the most important differences is the high degree of motion present in swarm systems. The robots are able to move continuously and rapidly through the environment in a way that the connections between the robots may be lost and may have to be reconstructed in different configurations. Due to this fact, groups of synchronized robots may be separated into several groups while moving and also different moving groups may meet somewhere in the environment to reconnect to each other, by this constructing a single group. Such problems of groups' separations and connections are not considered in the traditional time synchronization protocols developed for static wireless sensor networks.

The proposed approach is a kind of an overlay strategy which can be used with any of the time synchronization protocols for wireless communications. The purpose of the developed strategies is to handle the separation and reconnection problems which very often can be seen in swarm robotics systems.

4.2.1 *Swarm Separation*

Swarm robotics is a kind of continuously and rapidly moving network. where a high degree of motion can result in a lot of disconnections and reconnections resulting in new configurations concerning the links between the different nodes of the network which are moving around. Time synchronization, in most of the well-known protocols, is performed through a specific node referred to as *synchronization master*. The synchronization master is supposed to hold the reference time used to synchronize other nodes in the network periodically. All nodes which construct a connected network will be synchronized through the synchronization master. In networks where several connected groups of nodes exist and there is no connection between these groups, several masters, one for each connected group is needed. Swarm robotic systems belong to the described networks.

Applying time synchronization on swarm robotics implies the need of selecting a synchronization master for each group deployed on some location in the environment. As long as the robots are staying on their deployment locations, their master will synchronize them periodically. However, at some point of time the robots will start moving in the environment. The links between the robots, which were within one group, may become disconnected and new links with new robots will appear due to the motion. Due to these facts robots being synchronized through some defined master will lose their master of synchronization. The previous master will stay within one of the resulting sub-groups. Consequently the other separated groups will continue moving without having any master to synchronize with. Over the time, the robot clocks will drift from each other and the robots will become increasingly non-synchronized.

In applications, where robots are performing a cooperative task, which needs synchronization, a strategy is necessary to work on the continuous separations in the swarm caused by the robots' movement. The strategy goal then will be to select a synchronization master for each group, as soon as it loses its master. Guaranteeing a synchronization master in each connected group, will keep the robots of the group synchronized to each other and is necessary for selecting a shared master when different groups reconnect to each other while moving as we will see later.

The kind of separation we are interested in is group separation. Group separation is a likely situation in swarm robotics according to, for example, the assignment of different tasks while robots are moving. Another example is a swarm separated into two ones while some of its robots where applying obstacle avoidance and get disconnected from the others. Group separation indicates the situation where a group of robots separates from its original group and starts to move in the environment alone. Our approach guarantees selecting a master for synchronizing each separated group. Selecting a robot in the separated group can be done simply according to some ID selecting algorithm just to select the robot with the lowest or with the highest ID to be the master. However our strategy is to select the robot which has the smallest drift to the previous synchronization master of the original group. Selecting the robot which has the smallest drift will assure that the separated groups will preserve a similar time scale after separating.

4.2.1.1 Master Selection for Separated Groups

Normally time synchronization is applied on all connected nodes in the network according to some hierarchy. The synchronization error increases with increasing

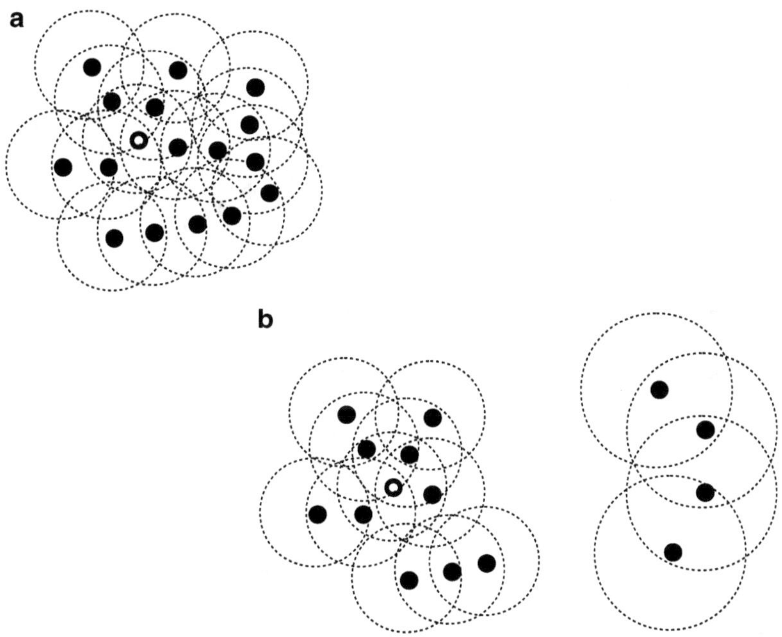

Fig. 4.1 (**a**) Minimum communication range within the swarm (**b**) Maximum communication range during separation

the level on the hierarchy as it is accumulative over the hierarchy levels due to the wireless communication delays. The robot, which has the smallest drift in comparison to its previous master in the original group, is the robot which was on the shortest hop-distance to the previous master. Our strategy will elect the robot in the separated group which had the nearest level to the previous master. Using the well-known algorithms for exchanging robot's level among the robots in order to select this one with the smallest level is actually a communication intensive action. Our strategy is based on local timers of the robots which are set as soon as they have been separated from their groups. The duration of the local timer will be weighted based on the robot's level, i.e. the timer duration is related proportional to the level of the robot in its current hierarchy. As a result, the timer of the robot, which was on the nearest level to the previous master, will expire before other timers. As soon as the local time of the robot expires, the robot starts to broadcast its level to all robots within its communication range. All robots receiving the message and finding out that the included level is higher than their level do not broadcast their level and accept the sender as their new synchronization master. It is assumed that each robot uses a limited communication range while exchanging time synchronization messages within its connected group to save energy and avoid collisions as in Fig. 4.1a. However, when a separated robot broadcasts its level to other separated ones, it will use its maximum communication range for this particular broadcasting. Using the maximum communication range in this case will increase the probability for the broadcast to reach as much separated robots as possible, see Fig. 4.1b.

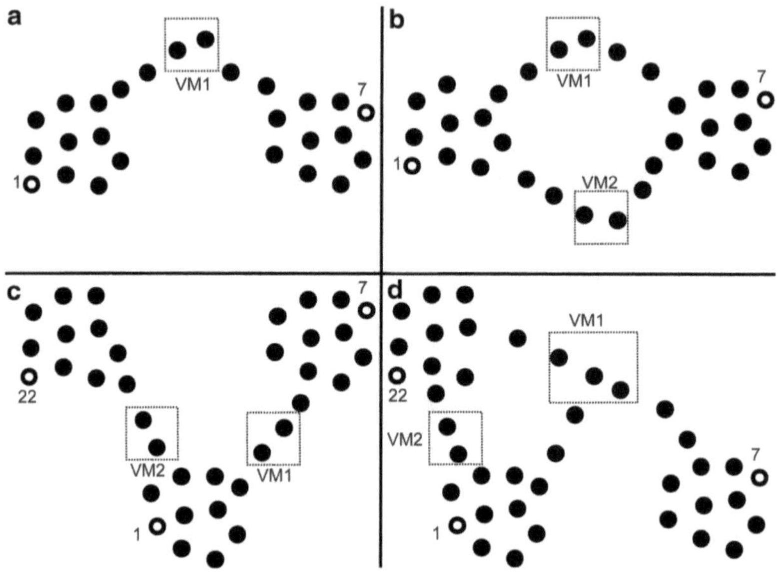

Fig. 4.2 (**a**) Single connection between two swarms (**b**) Multiple connection between two swarms (**c**) Multiple connection between many swarms (**d**) Multiple connection between many swarms with an inclusion relationship

4.2.2 Swarm Connection

Connections between different swarms of robots are mostly to appear in swarm robotics according to the continuous motion of the robots in the environment while performing different tasks. The connected groups may own different masters for synchronization, which could have synchronized the groups for a considerable duration of time. Consequently, the different groups may have different time scales according mainly to the different delays of the wireless communications within the different groups while exchanging time synchronization messages and due to different clock drifts on the master clocks. Let us assume now that the groups are going to participate on a shared cooperative task where time synchronization among the robots is a requirement, then a common time scale will be needed on all robots in the connected groups.

Group connections can happen in many forms and as the movement in the environment is a non-restricted and fast one, there may be several configurations how group connections can appear. Our approach differentiates between the following cases of connections:

- Single connection: robots belonging to different groups get connected in one single point, as in Fig. 4.2a.

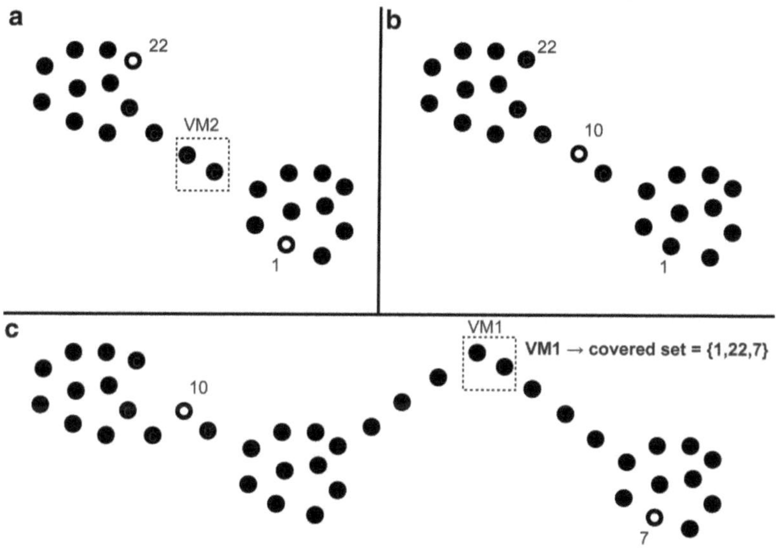

Fig. 4.3 Illustration of master/virtual master covered set

- Multiple connections: robots belonging to different groups get connected in more than one point. The different points of connections can connect the same set of groups like in Fig. 4.2b or even different sets of groups like in Fig. 4.2c, d.

The goal of our approach is to handle all possible cases of groups' connections in order to guarantee obtaining a common master for all connected groups which in turn guarantees having a single time scale on all robots. Before going through the approaches, some terms are essentially to be defined:

- Swarm ID: In our approach, each connected set of robots is referred to as a swarm. Swarms are defined though unique IDs, which are the IDs of robots playing the role of the synchronization master in the swarms. For example, the ID of the left swarm in Fig. 4.3a is 1, as the robot playing the master role is the robot with $ID = 1$.
- Virtual Master VM: In the phase of connection a group of connected robots originate from different swarms. A virtual master represents the robots from different swarms and different time scales, which meet in the communication range of each other while moving. One robot out of them will be selected later according to our approach to become the synchronization master of the final group consisting of all connected moving swarms.
- Master Covered Set: This is a set of swarm IDs belonging to swarms, which up to now have been covered by the current master. It is possible for a virtual master exactly like for a master to have a covered set. In case of master's covered set, it is the set of swarms which are covered up to now by the current master like

in Fig. 4.3b. However for virtual master, the covered set is the set of swarm IDs belonging to swarms which are covered by all masters of the connected groups, like in Fig. 4.3c.

The developed approach relays on being able to handle the problem of selecting a shared master in order to guarantee the same time scale for all robots. Such a problem is able to be solved using some well-known selection algorithm like smallest ID selection. However, using any of these well-known strategies will be on the cost of communication complexity. In our work we aim to select the most suitable master at as low communication complexity as possible.

All previously developed protocols for time synchronization in wireless communications use the message exchange concept to send and receive timestamps necessary to synchronize network nodes. Our approach uses exactly these messages without adding any additional communication complexity. Let us assume any of the well-known time synchronization protocols selecting a master node starts sending time synchronization messages periodically to other nodes in the network. According to the developed approach the message includes additionally:

- Current master ID: The *ID* of the current master which is synchronizing this specific node.
- Covered set: The set of groups covered by the current master of the node.

Based on this information, the recognition of connection situations between different groups will be possible.

4.2.2.1 Recognition of Swarm Connections and Virtual Masters

Let us assume that the swarms in the environment apply some kind of periodic time synchronization through a master robot. While moving around, a robot from group-*A* can move into the communication range of another robot from another group-*B*. Each time the robot gets a periodic synchronization message from its master it checks the two received entities of information against the locally stored ones, which are the master ID and the covered set of the master. Both entities of information are sent when the connection happens between group-*A* and group-*B* as part of the synchronization message as mentioned above. At some time, the robot from group-*A* connected to group-*B*, will receive the periodic synchronization message from the master of group-*B*. As soon as the group-*A* robot receives the synchronization message, it will compare the information included in the message with the locally stored information. The result of the comparison will be the detection of a connection between different swarm groups. The same result will be drawn out on the robot from group-*B* which is connected to group-*A*. Now each robot needs to decide whether to accept the synchronization message from the other master or to reject it or even to select a new master. The decision will be according to the kind of connection which has happened.

4.2.2.2 Virtual Master Selection

A virtual master situation is a situation, where different robots belonging to different groups are connected to each other due to the movement of the swarm robots in the environment. The connection can happen in different configurations which are mentioned above. We need to distinguish between the following cases:

- Single virtual master: There exists a single virtual master situation like in Fig. 4.2a: In this case the robot in the virtual master with the smallest ID will be selected as the new master for the connected groups. The communication in this case will be limited to be between the robots in the virtual master and not between all robots in the connected groups like it would be if we would apply the *ID* selection strategy.
- Multiple virtual masters: There exist multiple virtual master situations. The covered sets of the different masters can be equal to each other like in Fig. 4.2b. Here the covered set of $VM1$ and $VM2$ is $\{1, 7\}$. At the same time, it is also possible to have the covered sets of the virtual masters not to be equal to each other. In this case there are two possibilities: the first possibility is that the covered sets of the different virtual master are related to each other through an inclusion relationship like in Fig. 4.2d where the covered set of $VM2$ is $\{1, 22\}$ and the covered set of $VM1$ is $\{1, 7, 22\}$. The second possibility is to have the covered sets of the virtual masters intersected with each other but not related with an inclusion relationship, like in Fig. 4.2c. Here the covered set of $VM1$ is $\{1,7\}$ while the covered set of $VM2$ is $\{1,22\}$. In both cases, when multiple virtual masters exist, the first step is to select one of the virtual masters and after this to select one robot from the selected virtual master to synchronize the robots in the connected groups. Selecting one of the existing virtual masters will be the job to be done by the current masters of the connected swarms. After that, the selection of the robot from the already elected virtual master will be done through communication between the robots within the elected virtual master themselves.

The selection of virtual masters is based on the time point at which the existing virtual masters are announced to the current masters of the swarms. Here we can distinguish between two possibilities:

- Different virtual masters announce times: This possibility has a very high probability as in most cases the current masters of the connected groups will not be informed about the multiple virtual masters at the same time instance. This high probability is based on two facts: The mostly different appearance times of virtual masters in addition to the different distances on one hand and communication delays between robots in virtual masters and the masters of their groups on the other hand. This will lead to different time points of announcing the virtual masters to the current masters of the connected groups. Due to this fact, the firstly announced virtual master will be selected by the connected groups. After that one robot from the selected virtual master will be chosen to synchronize the connected groups. In case that another virtual master is

announced later, the current master of the resulted group will check the covered set of the newly announced virtual master and compare it to its own one. If the newly announced virtual master is covering exactly the groups covered by the current master, or a set of groups which is included in the set covered by the current master, then the newly announced virtual master will be rejected and its robots will be informed to attend themselves to the current existing master. However, if the set of groups covered by the recent announced virtual master is larger than the covered set of the current master, the new virtual master will be selected and the synchronization master will be elected from its robots.
- Simultaneously virtual master announcements: The possibility that the masters of the connected groups get informed about the existence of multiple virtual masters simultaneously has a very low probability. However, when this happens, the covered sets of appeared virtual masters can be equal, or one of them includes all other covered sets. In this case, the virtual master which owns the covered set including the largest number of connected groups will be selected. On the other hand, it is possible that the appeared virtual masters' own covered sets are not related through an inclusion relationship and are covering different groups. However, the covered sets must share at least one covered swarm. Based on the fact that there must be at least one shared swarm between the appeared virtual masters. Its master will be informed about the different virtual masters. In this case, the master of the shared one will become the new synchronization master of all connected groups.

4.3 Case Study: Time Synchronization for Object Tracking

4.3.1 Scenario

The scenario we are considering in this chapter is area surveillance, (i.e. tracking mobile objects in a complex and dynamic environment). Such scenarios are mainly seen in military applications such as tracking enemy vehicles and monitoring the borders for illegal intrusion detection. It can also be found in control and safety applications or civil applications such as intruder tracking and wild animal tracking for studying their motion patterns and behaviors. Object tracking can be considered to have soft real-time behavior, where meeting the time requirements improves the performance of the system and missing them decreases it. In object tracking scenarios, to monitor movement patterns, and to register interesting events efficiently, robots need to have a common time scale. It allows the system to analyze the tracking results, build the motion patterns and order the events correctly in time.

The environment which we are considering changes dynamically and continuously. Hence, no infrastructure is available for monitoring and controlling. It is also inefficient to deploy a fixed network that supports tracking during the operation.

Hence we are interested in deploying robotic swarms to perform this task. Deploying mobile robots for tracking is beneficial, because they can cover a wider area over time and can re-position themselves for efficient tracking, in addition to their ability to cope with changes that can occur in a dynamic environment. In our work, we consider the scenario of single object tracking. However, the approaches we are using can easily be extended towards multi-object tracking, as the swarm-based approaches provide a high level of scalability through localized rules used in the algorithm design.

The environment we consider is a bounded 2-D area denoted as A. Swarms S of N_S robots each are available for tracking the moving object. Each robot R_i, has limited sensing capabilities which is used for identifying the target to be tracked. The sensing range of each robot is assumed to be a circle with radius r_s. The mobile object to be tracked is referred to as O. It is assumed that the total area covered by the visible sensing area of the robots is much less than A. Hence the robots have to move in order to track the object continuously. Robots are equipped with wireless transceivers for communicating with other robots. We assume the unit disk graph wireless model [1] for wireless communication, where each robot can communicate with others located within a circle of radius r_c. We also assume that the communication area of one robot is much less than A. Hence, the robots have to send packets over several hops to reach a particular destination (other robot or base station) and the network need not to be always connected. We also assume that the communication radius r_c is larger than the sensing radius r_s. When the target is within the sensing range of any robot, it is considered to be tracked. The objective of our tracking is to find motion control strategies so as to maximize the total time, the object being tracked. In order to support cooperative tracking by swarms of robots, the robots are making guesses concerning the estimated future location of the target based on its current movement pattern. The predictive tracking range defines the range in which targets localized by other robots can affect the current robot's movements.

Mini-robots such as Bebots [8], shown in Fig. 4.4 are suitable candidates for this mission. These robots are equipped with a camera for target detection, an infrared sensor ring for obstacle avoidance and Zigbee and Bluetooth for communication. The camera they support is not omni-directional. Hence they have to use suitable motion patterns or spin behaviors to provide full coverage within the sensing region. Another module they support is a Nanotron radio signaling system [8], which uses RF ranging techniques to estimate the position of the nearby robots. We assume that, using this positioning system each robot can estimate the position of all other robots within the communication range.

4.3.2 Object Tracking

For our tracking scenario we assume that the robots are deployed initially in different locations in the environment as separated swarms. Each swarm consists of

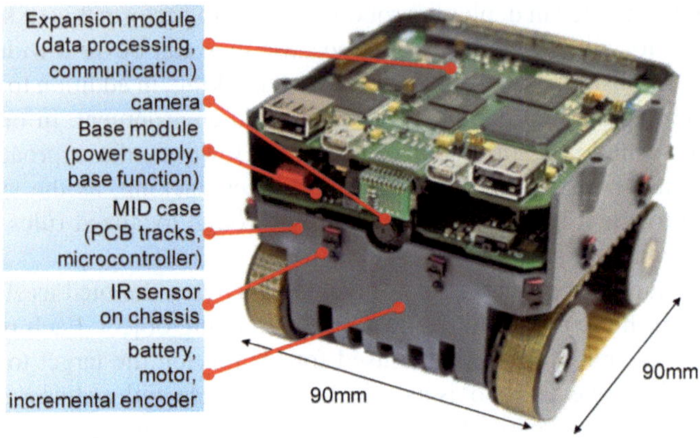

Fig. 4.4 Bebot mini-robot

a number of robots which are going to be used in tracking. In each of these swarms a master is selected. The master will apply a kind of periodically time synchronization on other robots in the swarm.

For our case study we have implemented a simplified version of the object tracking approach as presented in [19]. The Cooperative Multi-Robot Observation of Multiple Moving Targets (CMOMMT) uses a potential field-based approach for object tracking. Each robot defines three ranges: sensing range, predictive tracking range and communication range [19]. Robots detect targets, other robots or objects that are within the sensing range. The predictive tracking range defines the range in which targets localized by other robots can affect the current robot's movements. The communication range defines the region where they can send (receive) messages to (from) each other. It is usually larger than the sensing range and predictive tracking range.

A robot can know about two types of targets [19]: those that are directly sensed or those that are "virtually" sensed through predictive tracking. When a robot receives a message regarding the location and velocity of a sighted target that is within its predictive tracking range, it begins a predictive tracking of that target's location, assuming that the target will continue linearly from its current state. This predictive tracking will then give the robot information on the likely location of targets that are not directly sensed by the robot, so that the robot can be influenced not only by targets that are directly sensed, but also by targets that may soon enter the robot's sensing range.

The local control of a robot is based upon a summation of force vectors which are attractive for nearby targets and repulsive for nearby robots [19]. The robot is repelled from a target if it is too close to that target, and is strongly attracted to the target if it is nearby. Beyond a given distance, the attraction falls off linearly until the predicted tracking range is met, at which point the attraction goes to 0. The repulsive forces between robots are defined as: If the robots are too close together, they repel strongly, if they are far apart, they have no effect upon each other and when the distance between them is between these values, the magnitude of the repulsive force scales linearly [19]. The weighted local force vectors are combined to generate the commanded direction of robot movement.

4.3.3 Swarms Connection: The Need of Virtual Master Selection

As mentioned before, according to the tracking object strategy we are studying, as soon as the object we aim to track is detected by any robot in any of the deployed swarms, the robot will attract other robots in the swarm to start tracking the detected object. If the object is within predictive tracking range of at least one robot or a swarm, the whole swarm will start to move tracking the moving object. While moving, the master of the swarm will keep on synchronizing the other robots in its swarm. However, it will be very probable that the object we are tracking attracts other swarms located on different positions in the environment to follow it as well. Different swarms holding different time scales start tracking the object and are going to get connected to each other while moving. In order to be able to perform the tracking task correctly, it is necessary for all robots belonging to the union of the different swarms to have a common time scale to provide correct data fusion and tracking results. Applying our newly developed approach to select a shared master when different swarms get connected to each other enables them to handle the different possible cases of connection.

4.3.4 Swarm Separation: The need of Shared Master Selection

There are many reasons for a swarm tracking some moving object to get separated into different swarms. For example, a group of robots belonging to the tracking swarm may be prevented from being able to continue tracking the object due to facing some obstacle in the environment while moving. The separated group then needs to be kept synchronized in order to be able to re-track the object as soon as it is attracted again to track it. In such a situation there is a need of having a master for this separated group anyhow. In addition the existence of such a master is mandatory to allow the application of our virtual master concept in case of other tracking groups getting connected to this separated one.

4.4 Results

The newly developed approach has been evaluated using the well-known robotic simulator Webots. Webots is a robotic development environment which allows modeling, programming, and simulating robotics. It is a 3D simulation environment,

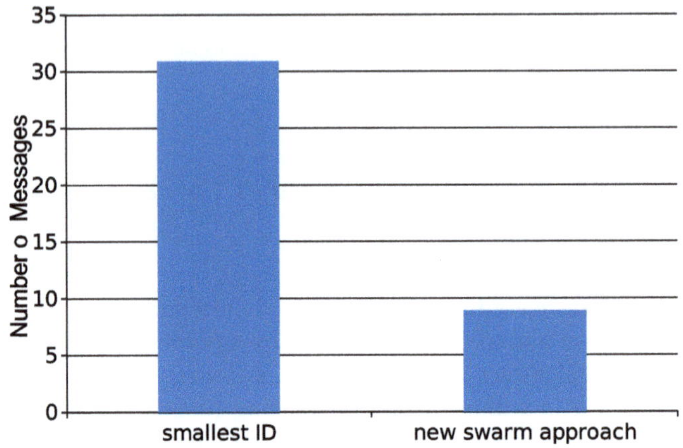

Fig. 4.5 Comparison of communication complexity of swarm separation and smallest-id selection strategies

which gives the possibility to model robots easily and to program with different languages like C++, java, python, etc.

Different scenarios were tested with modifying a number of parameters including the robots number in swarms, the tasks number and positions which need to be visited by the robots of swarms. First we have evaluated the synchronization master guarantee in case of swarm separation while moving. After that we have evaluated the second concept of the developed approach which involves the guarantee of selecting a common master for connected swarms moving in the environment with different configurations of connections covering the above discussed cases. In all evaluations the time synchronization protocol used is T-Sync [2].

4.4.1 Swarm Separations Evaluation

20 robots were deployed in the environment and one of them was elected to be the master to synchronize other robots using the T-Sync protocol. It is assumed after applying the time synchronization for some time that a new task will be assigned to 9 of the robots from the swarm which needs them to move to the position of the new task. Here we have applied the newly developed strategy to elect a master for synchronizing the separated group of nine robots which are elected in a random fashion from the different levels of the original hierarchy. The scenario was tested 30 times to allow different selection of the separated robots and their levels. The message complexity was measured and compared to those results from using the traditional ID selection algorithm to select a master in each time a separation occurs, see Fig. 4.5.

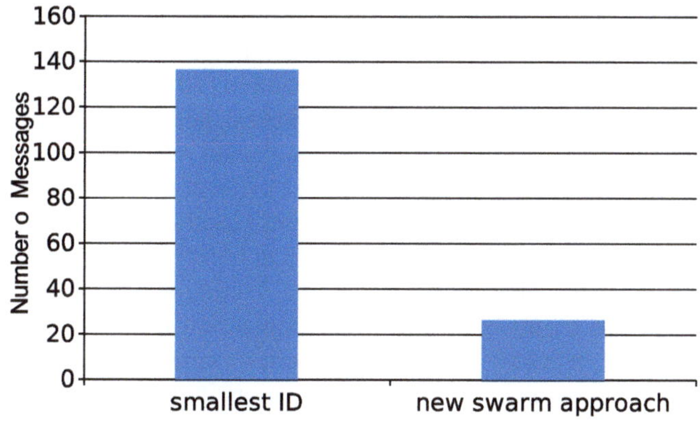

Fig. 4.6 Comparison of communication complexity of swarm connection and smallest-id selection strategies

4.4.2 Swarm Connections Evaluation

30 robots were used in this test scenario. The different robots are deployed initially in three locations in the environment. At their initial locations, a master is elected in each of these swarms to synchronize the other robots in the swarm using the T-Sync protocol. At some point of time a task appears in the environment and its location is communicated to the three swarms to move towards it and perform it. The moving swarms will then get connected to each other while trying to visit the task location.

The scenario was tested 30 times which has allowed different configurations of the swarm connections to appear. The new approach was applied to guarantee the selection of a common master to synchronize all the connected swarms in all configurations of connection. The message complexity resulted from selecting this master according to the new approach was compared to the one resulting from applying the well-known ID selection algorithm, see Fig. 4.6.

4.5 Related Work

Time synchronization is a well-studied problem in the literature on wireless communication. There is a set of protocols developed mainly for the clock synchronization purpose in wireless sensor networks. In wireless communication a set of path delays are added to the message exchange path between nodes and are accumulated over the multi hop communication in the network. Such delays do not exist in the wired communication. They include mainly, send delay, access delay, broadcast delay, and receive delay. These delays play the main role in the synchronization errors of wireless networks due to the non-determination and value range of most these delays. A description of these delays can be found in [20].

The time synchronization protocols for wireless networks can be categorized into two large families: sender–receiver protocols and receiver–receiver protocols. Sender-receiver approaches are well applied in the wired networks and have been imported to the wireless ones. These approaches based on the round-trip message concept to exchange timestamps between nodes in order to synchronize their clocks together. Examples for this approach are the Timing-sync Protocol for Sensor Networks [6] or the Lightweight time synchronization for sensor networks [23]. On the other hand, receiver–receiver approaches exploit the broadcast property available exclusively in the wireless communication, in broadcasting a time beacon by a reference node to other nodes within its communication range. The assumption made then is that all received nodes are going to receive the time beacon at the same time. This assumption allows avoiding the estimation of wireless communication delays. The beacon receiving event will then be used as a reference point of time to be synchronized. The most interesting receiver–receiver protocol used in wireless sensor networks is the reference broadcast synchronization protocol (RBS) [5].

Time synchronization was also studied for ad-hoc networks where the classical time synchronization algorithms are not applicable according to the sparse nature of these networks [21]. The main contribution of this study was providing a translation technique for the time stamps when they are exchanged between nodes in order to be able to order the events happened correctly and cope with possible disconnection in the ad-hoc networks.

Some studies are done already in the domain of cooperation of swarm robots. In [17] a kind of cooperation in terms of physical connections is exploited to pass a large gap impossible for one robot to be done. Another example of physical cooperation among swarm robots is used for object transportation tasks in [7]. The cooperation also among a heterogeneous swarm of robots is studied in [4] where wheeled robots cooperating with flying ones in navigating and finding efficient paths through a cluttered environment.

Object tracking as a cooperative task is a well-studied problem in wireless sensor networks and a large spectrum of literature is available [13, 14, 25]. Existing multi-robot tracking algorithms differ in the approaches used, some are centralized algorithms while others are distributed ones. One of the earliest work in distributed control algorithms for target tracking is presented in [18, 19]. We used the potential field based control algorithm A-CMOMMT in our case study. Other approaches for object tracking in multi-robot scenario is presented in [10, 12, 15, 24].

Several works on the practical aspects of object tracking such as sensor fusion, vision based tracking, prediction of target movement for collision avoidance in path planning have been published [16, 22]. Hybrid approaches which use stationary and mobile sensors for target tracking problem exist as well [9].

4.6 Conclusion

We have presented through this work a novel approach for employing swarm robotic systems in real-time self-organized tasks. Our study case was the object tracking task. The contribution of this work was mainly on extending existing time synchronization protocols for wireless communication and developing new strategies which allow using the well-known time synchronization protocols of wireless communication in swarm robotic systems. Two main strategies were developed in order to handle two main phenomenon resulted from the high degree of motion in the swarms which are: the separations and the connections. The strategies were also implemented and tested using the robotic simulator Webots, where the results were compares to one of the state of the art possible to use algorithm. The object tracking strategy used in our study case is the state of the art one: The Cooperative Multi-Robot Observation of Multiple Moving Targets (CMOMMT) uses a potential field-based approach for object tracking.

References

1. Clark BN, Colbourn CJ, Johnson DS (1990) Unit disk graphs. Discrete Math 86(1–3):165–177
2. Dai H, Han R (2004) Tsync: a lightweight bidirectional time synchronization service for wireless sensor networks. SIGMOBILE Mob Comput Comm Rev 8:125–139
3. Dressler F (2007) Self-organization in sensor and actor networks. Wiley, New York
4. Ducatelle F, Caro GAD, Gambardella LM (2010) Cooperative stigmergic navigation in a heterogeneous robotic swarm. In: Proceedings of the 11th international conference on simulation of adaptive behavior (SAB), Lecture Notes in Computer Science 6226, ISBN 978-3-642-15192-7, Berlin, Springer, pp 607–617
5. Elson J, Girod L, Estrin D (2002) Fine-grained network time synchronization using reference broadcasts. SIGOPS Oper Syst Rev Lecture Notes in Computer Science 6226 Springer 2010, ISBN 978-3-642-15192-7 36:147–163
6. Ganeriwal S, Kumar R, Srivastava MB (2003) Timing-sync protocol for sensor networks. In: Proceedings of the 1st international conference on embedded networked sensor systems, SenSys '03. ACM, New York, pp 138–149
7. Gross R, Dorigo M (2003) Evolving a cooperative transport behavior for two simple robots. In Artificial Evolution—6th Inter-national Conference, Evolution Artificielle, EA 2003, P. Liardet, P. Collet, C. Fonlupt, E. Lutton, and M. Schoenauer (Eds.), Lecture Notes in Computer Science, Springer Verlag, Berlin, Germany, 2936 :305–317
8. Herbrechtsmeier S, Witkowski U, RÃijckert U (2009) Bebot: A modular mobile miniature robot platform supporting hardware reconfiguration and multi-standard communication. In: Progress in robotics. Communications in computer and information science, vol 44. Springer, Berlin, pp 346–356
9. Jung B, Sukhatme GS (2001) Cooperative tracking using mobile robots and environment-embedded, networked sensors. In: Proceedings of IEEE international symposium on computational intelligence in robotics and automation, pp 206–211. DOI 10.1109/CIRA.2001.1013198
10. Jung B, Sukhatme G (2006) Cooperative multi-robot target tracking. In: Distributed autonomous robotic systems. Springer, Japan, pp 81–90

11. Khaluf Y, Mathews E, Rammig F (2011) Self-organized cooperation in swarm robotics. In: 14th IEEE international symposium on object/component/service-oriented real-time distributed computing workshops (ISORCW), Institute of Electrical and Electronics Engineers (IEEE), pp 217–226
12. Kolling A, Carpin S (2007) Cooperative observation of multiple moving targets: an algorithm and its formalization. Int J Robot Res 26:935–953
13. Kung H, Vlah D (2003) Efficient location tracking using sensor networks. In: Wireless communications and networking, 2003. WCNC 2003. 2003 IEEE. IEEE Computer Society, Silver Spring, pp 1954–1961
14. Li D, Wong KD, Hu YH, Sayeed AM (2002) Detection, classification and tracking of targets in distributed sensor networks. In: IEEE signal processing magazine, IEEE Signal Processing Society, pp 17–29
15. Liu Z Jr, Ang MHA, Seah WKG (2003) A potential field based approach for multi-robot tracking of multiple moving targets. In: The first humanoid, nanotechnology, information technology, communication and control environment and management (HNICEM)
16. Mazo M Jr, Speranzon A, Johansson K, Hu X (2004) Multi-robot tracking of a moving object using directional sensors. In: Proceedings of the ICRA '04. 2004 IEEE international conference on robotics and automation, vol 2, pp 1103–1108. DOI 10.1109/ROBOT.2004.1307972
17. Mondada F, Bonani M, Magnenat S, Guignard A, Floreano D, Groen F, Amato N, Bonari A, Yoshida E, Kröse B (2004) Physical connections and cooperation in swarm robotics. In: 8th conference on intelligent autonomous systems (IAS8), IOS Press, pp 53–60
18. Parker L (1999) Cooperative robotics for multi-target observation. Intell Automat Soft Comput 5:5–19
19. Parker L, Emmons B (1997) Cooperative multi-robot observation of multiple moving targets. In: IEEE international conference on robotics and automation, Institute of Electrical and Electronics Engineers (IEEE), vol 3, pp 2082–2089
20. Ping S (2003) Delay measurement time synchronization for wireless sensor networks. Technical report, Intel Research, Intel Corporation
21. Römer K (2001) Time synchronization in ad hoc networks. In: Proceedings of the 2nd ACM international symposium on Mobile ad hoc networking & computing, MobiHoc '01. ACM, New York, pp 173–182
22. Schulz D, Burgard W, Fox D, Cremers AB (2003) People tracking with a mobile robot using sample-based joint probabilistic data association filters. Int J Robot Res 22:2003
23. Van Greunen J, Rabaey J (2003) Lightweight time synchronization for sensor networks. In: Proceedings of the 2nd ACM international conference on wireless sensor networks and applications, WSNA '03. ACM, New York, pp 11–19
24. Werger BB, Mataric MJ (2000) Broadcast of local eligibility for multi-target observation. In: Proceedings of distributed autonomous robotic systems. Springer, Berlin, pp 347–356
25. Xu Y, Winter J, Lee WC (2004) Prediction-based strategies for energy saving in object tracking sensor networks. In: IEEE international conference on mobile data management, Institute of Electrical and Electronics Engineers (IEEE), p 346

Chapter 5
Improving Performance of Controller Area Network (CAN) by Adaptive Message Scheduling

Tobias Ziermann, Zoran Salcic, and Jürgen Teich

Abstract Controller Area Network (CAN) has been widely used in many distributed control applications. The low capacity of CAN and usually strict constraints under which it is used in real-time applications limit its applicability. This paper presents a new approach for scheduling messages on CAN based on online adaptation of message offsets, which results in significantly shorter average message response times compared with approaches which guarantee response times. This allows much higher utilization of CAN, resulting in increased effective capacity, and enables its use in applications in which more expensive faster protocols would be required otherwise. Since the average message response times are reduced, the approach is particularly attractive for soft real-time applications. In addition, the paper proposes a method for self-initialization of the CAN-based system on its start-up, which reduces response times from the very start of the system. The proposed approach is very simple and requires modest resources for its implementation, with no change of the standard CAN infrastructure. We demonstrate the performance improvements by comparing the approach with other approaches.

5.1 Introduction

Distributed control systems use special kinds of networks, called buses, for communication between their components, which are usually called *Electronic Control Units* (ECUs) or nodes. The most common communication bus is the *Controller*

T. Ziermann (✉) • J. Teich
Department of Computer Science, University of Erlangen-Nuremberg, Germany
e-mail: tobias.ziermann@cs.fau.de; teich@cs.fau.de

Z. Salcic
Department of Electrical and Computer Engineering, The University of Auckland, New Zealand
e-mail: z.salcic@auckland.ac.nz

M.T. Higuera-Toledano et al. (eds.), *Self-Organization in Embedded Real-Time Systems*, DOI 10.1007/978-1-4614-1969-3_5,

Area Network (CAN) [6]. The ECUs connected to CAN are typically microcomputers, which execute parts of the application. Examples of such distributed control systems are in automotive, industrial control, medical and other similar applications [3, 5, 8, 12, 16, 19].

Compared with general purpose computer networks, industrial communication buses, such as CAN, typically have fairly low nominal transmission speed and bandwidth, but also very short response times. However, when multiple messages are sent at the same time, the messages with low priorities can suffer from long response times, sometimes leading to starvation. This problem becomes more severe with increasing traffic on the bus. For example, more than 70 ECUs are used in a distributed embedded system of a modern car [17]. Because the underlying CAN bus has limited bandwidth and utilization of the bus is high, this inevitably results in increased message response times. In order to keep the message response times within specified limits, the bus has to be run at low bus utilization and thus reduced likelihood of simultaneous message releases from different nodes. A possible solution for a given distributed system and a set of nodes would be to increase the bandwidth by using another communication bus or by using multiple CAN segments at higher cost. In this chapter, we propose to retain the existing CAN infrastructure and add the intelligence to the nodes that will result in better scheduling of messages and shortening of the average message response times, while at the same time allowing higher utilization of the bus. The end result is achieved with a small penalty in the form of monitoring of the traffic on CAN bus by all nodes and executing an adaptive scheduling algorithm that changes the release times of messages sent by individual nodes.

The proposed approach for message scheduling specifically targets soft real-time systems and is not intended for hard real-time systems. As soft real-time we consider applications where satisfying timing constraints is important for the system functionality, but missing the deadlines once in a while will not result in malfunction of the system. Typical examples of such applications are found in the body network of an automotive embedded system [17]. For example, a rain sensitive wiper needs to react in time on the changing environment, but some delayed messages will not cause a system failure.

Current analytical techniques for calculating the message response times for priority-based buses are overly pessimistic [7, 14], and result in very conservative designs with low bus utilization. On the other end of the spectrum are approaches that use simulation models of the behavior and measure the message response times [11, 21, 24]. These approaches are based on the assumption of random release times of messages by the system nodes, which may, in the worst case, result in simultaneous release of messages by multiple or all nodes. Our approach belongs to the later group, but is based on the distribution and dispersion of message release times, thus alleviating the problem of simultaneously released messages. Assuming that messages released by different nodes are periodic, offsets are added to the release times using an offset adaptation algorithm that runs on each node without change of the message format and without any modification of the CAN bus infrastructure. These offsets cannot be calculated in advance because of the asynchronous nature of the distributed system, where nodes are driven by different clocks and with no system-wide reference clock. From this reason our approach is based on dynamic adaptation of offsets (thus called *dynamic offset adaptation algorithm* (DynOAA) [26]), where each node that releases messages monitors the traffic on the bus and adapts the offsets as the traffic changes. DynOAA achieves significant reduction of the average of maximum message response times, and enables the use of the same bus at higher bus loads. Compared to our initial work [26], this chapter presents the following contributions: (1) The original approach is extended to the initialization mode of the system, so the system can autonomously pass from initialization to normal mode in which DynOAA is applied. (2) Qualitative criteria to compare CAN scheduling techniques are defined and used to compare our approach to others. (3) Simulations are presented that show the quantitative results of our approach.

This chapter is organized as follows. In Sect. 5.2, we position our approach in the context of related work. Section 5.3 defines the problem by describing the system model and the evaluation criterion. Section 5.4 gives the details of the proposed dynamic adaptation method (DynOAA) and illustrates its operation on examples. Section 5.5 evaluates our approach by comparing qualitatively the properties and quantitatively the results with related works. Conclusions and future work are given in Sect. 5.6.

5.2 Related Work

A trend towards adding more nodes to the distributed embedded and control systems applications faces the limited bandwidth and message response times as the major obstacles. An obvious solution to this problem is to try using faster buses such as FlexRay [10] or Ethernet [9]. However, the introduction of new bus technology is linked with a huge cost, because it requires the change of not only the bus infrastructure, but also modifications of both hardware and software on the ECUs. Furthermore, the faster communication bus protocols use different arbitration methods. For example, FlexRay uses two access mechanisms. First, a priority-based event-triggered access in a-priory defined time slots (dynamic segment) is possible. Second, a static time-triggered access is offered (static segment), where the schedules are fixed at planning time. The dynamic segment is scheduled similar to token passing. An order of the different applications is specified offline. Following this order, each application can either transmit or pass the token. At the beginning of the next dynamic segment it starts again with the first one. This approach has larger response times than CAN because (1) the messages have to wait for the dynamic segment to start, (2) for the token to be passed, and (3) possibly even to wait for several segments, because the token did not reach the application before the segment completed.

The static segment has the typical properties of a time-triggered access mechanism: Periodic time slots are reserved statically at design time. This has the advantage that the response-times for ideal periodic applications are deterministic, given that the application triggering the message is synchronized with the communication. But for applications that are not active for the full system operation time the disadvantage is that the bandwidth is wasted when the application does not require the time slot. Additionally, the time-triggered access mechanism cannot easily accommodate the system changes during its development and it always has to be planned with complete system knowledge, which is not always available and possible. Comparison of time-triggered and event triggered approaches has been subject of many discussions [1] and there is a sound argument of the advantage of using time-triggered approach for systems with purely periodic message release. However, as many existing and future systems still rely on use of traditional buses, in our case CAN, the focus of many researchers has been how to better utilize CAN and effectively extend its applicability and life. Besides retaining legacy systems and huge investment into various kinds of nodes and applications, CAN still offers its original advantages of simple implementation, short response times and ease of change of the designed system. In some applications, e.g. automotive, the applicability of CAN has been extended by using several bus segments, which accommodate smaller number of nodes, interconnected through a gateway [22]. However, this approach results in increased response times, because the gateway itself introduces additional delays. Moreover, analyzing the timing behavior of such systems is more difficult, because these delays have to be estimated and taken into account in system design.

There has been a significant body of research on how to allow a time-triggered access on CAN by using new protocols on top of CAN. In that case, a single node organizes the scheduling of the messages, e.g. the FTT-CAN Protocol [2] or the server-based communication [18]. Even though the guaranteed response times are improved, the disadvantages of these approaches are that the scheduling node (master) needs to send extra scheduling messages and all other nodes (slaves) need to register their transmission. This causes additional traffic over the bus that further reduces the already small bandwidth. Another disadvantage is that all nodes need to follow this protocol, making incremental changes in system design and integration difficult. The approach proposed in this paper addresses the aforementioned disadvantages by using a fully distributed approach to message scheduling without introducing additional communication overhead. Guaranteeing response times is also possible by using an analytical approach and taking advantage of the fact that several message streams are on the same ECU. Fixed offsets between the message streams on one node are assigned by using an offline heuristic algorithm [12]. In the remaining text, we refer to it as to static offset assignment algorithm (staticOAA). The basic idea of the staticOAA is to spread the release times of all messages as far as possible. This is achieved through creation of an array of possible start time slots within the length of the largest message release period for each node. The message streams are ordered according to increasing period. For each stream the following steps are performed:

1. Find the longest interval in which there is no load on the bus,
2. Set the message release of the stream in the middle of this interval and

3. Update the array with all messages of this stream. According to the resulting schedule, the offsets are assigned for each stream.

The authors have shown that the Worst Case Response Times (WCRTs) for the streams are reduced compared to the worst case scenario in which all offsets are zero. Assigning offsets to the messages results in better response times, but takes the asynchronous nature of the nodes that communicate over bus only into account by considering worst case assumptions. The approach also does not take into account the dynamics of the system and resulting variations in the bus traffic. Instead, everything is based on a-priori given, static assumptions. In contrast to the staticOAA, the approach proposed in this paper is based on monitoring of the current network traffic and adapting message offsets in order to reduce the likelihood of the simultaneous message releases on different nodes.

5.3 Problem Definition and Performance Measure

In order to define the problem and present our solution, some basic descriptions of the underlying CAN protocol are provided first. The CAN specification [6] defines the data link layer and roughly describes the physical layer of the ISO/OSI-Model. The physical setup consists of a terminated 2-wire bus. Bits are represented by the Non-Return-To-Zero method, where a logical "0" (dominant state) is represented when there is a voltage difference on the wires, and a logical "1" (recessive state) represented with equal voltage levels on the wires. This bit representation results in a dominance of the logical "0". The length of a bit depends on the bit rate used, e.g. 1 μs at 1 Mbit/s.

In data link layer, a message oriented approach is chosen. Four different types of frames are used to transfer messages: Data frames, Remote Transmit Request frames, Overload frames and Error frames. The most important and most frequent is the data frame, which is used for data exchange. In this work, we will only consider the data frames. The analysis of the influence of the other types of frames is left for future work. Each data frame has its unique identifier, which at the same time defines the message priority used when requesting the bus access. Bus arbitration is done by *Carrier Sense Multiple Access with Bitwise Arbitration* (CSMA/BA) method, which works as follows: Each node that attempts to access the bus starts transmitting its message as soon as the bus is idle for 3 bit time slots. At the same time, the node monitors if any other node is transmitting. When the transmitted bit differs from the one received on the bus, then a message with higher priority is also being sent and the node stops its own transmission. After transmitting the identifier, only the message with the highest priority is left and gets exclusive bus access. In this section, we first introduce the system model, as well as the rating function which will be used in evaluation of performance of the proposed approach and its comparison with the other approaches.

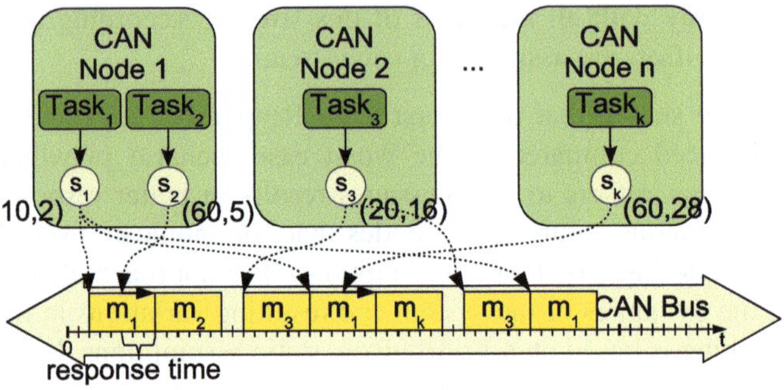

Fig. 5.1 System model

5.3.1 System Model

The CAN system we target can be described by a set of nodes communicating over the bus as shown in Fig. 5.1. One or more tasks on each node initiate the communication, typically periodically, i.e., they release messages. However, while there is a common time reference for tasks on a single node, there is no common time reference for the overall system, thus the system is asynchronous. It is important to note that the meaning of asynchronous in this paper differs from one used in some scheduling related literature [23], where synchronous is referred to tasks with zero offsets and asynchronous refers to tasks with non-zero offsets.

In our model, we abstract the tasks by considering only the mechanism used to release messages called a stream. A stream s_i is characterized by a tuple (C_i, T_i, O_i) with $0 \leq O_i \leq T_i$, that is, by a transmission time C_i, by a period T_i (time between any two consecutive messages generated by stream s_i) and an offset O_i. The offset is relative to a global time reference. It can therefore drift over time, because the local time reference differs from the global one. The hyper-period P is the least common multiple of all periods $lcm\{T_1, T_2, \ldots, T_k\}$. Assuming a synchronous system, the schedule is finally periodic with the period equal to the hyper-period. A scenario we consider consists of k streams, as illustrated in Fig. 5.1, where the numbers in brackets describe the period and the offset of each stream. The bottom part illustrates the behavior of the messages over time. The priority by which access is granted to the bus is given by the numbering of the streams, where a stream with a lower index, e.g. stream s_i, always has higher priority than a stream with a higher index, e.g., stream s_{i+1}. We assume the priorities are given by the designer who usually sets the priorities according to the stream period, so that a rate monotonic scheduling is achieved. A message is a single release or CAN frame of the stream. The time between a message release and the start of its uninterrupted transfer over the bus is the response time of the message. We do not add the constant non-preemptive time to transfer the message to the response time, thus a zero response time is always the best possible case for any message regardless of its length. This simplifies comparisons between different schedules. For example, the response time of message m_2 in Fig. 5.1 is three time slots, because it is delayed by the message m_1, which is being sent. The worst case response time of a stream s_i during a fixed time interval starting at time t_a and ending at time t_b (i.e., $WCRT_i(t_a,t_b)$) is the largest response time of the messages of the stream recorded during that time interval. An analytical approach calculates $WCRT_i(0,\infty)$.

In our approach, we assume discrete model of time with a minimal time resolution defined in advance. All stream characteristic times are multiples of this minimal time resolution. As mentioned before, in an asynchronous system the offset for all the streams can take any value which is an integer multiple of the minimal time resolution. In our approach to calculating offsets we assume that the individual streams are not bound by any constraints, but rather they can freely be set by the designer [13]. Finally, because our scheduling considers the entire distributed system that includes all message generating streams, the fact that one or more streams are on the same node does not make any difference. Therefore, in our analysis we assume that there is only one stream per node, without affecting the generality of the presented results. Also, as in the most of the other related approaches we do not take into account errors on the CAN bus, effectively assuming that ideal data transfers over bus.

5.3.2 Rating of the Schedule

In real-time systems task scheduling, the quality criterion of a schedule is typically based on a measure of whether or how many deadlines are met. This metric is inadequate for many practical problems because it is not so easy to define a deadline. If we look, for example, at a classical control application, a sensor-controller-actuator chain is triggered periodically and therefore fits to the task model. The problem with control algorithms is that one cannot decide beforehand what maximum delay and jitter the algorithm can tolerate, especially because the delay and jitter may increase over time, and the deadlines cannot be provided in advance.

Another method to evaluate the system performance is to compare WCRTs of all streams as in [12]. An example for an automotive scenario is presented in Fig. 5.2. The details of scenario 1 are provided in Sect. 5.5.2. The fully asynchronous case (1) assumes that all streams are asynchronous and, therefore, can all occur simultaneously. The second case assumes that the streams on each node are synchronized and the offsets are determined by staticOAA. In contrast to the first two approaches, in the third and fourth approach response times are not guaranteed WCRTs but correspond to maximum and average of response times found in large number (in this case 1,000) different simulation runs with random uniformly distributed offsets. Here, the assumption that all streams are synchronous is used, i.e. the offsets do not change. This assumption is considered reasonable, as the different clocks drift slowly and the offsets stay unchanged for rather long time intervals.

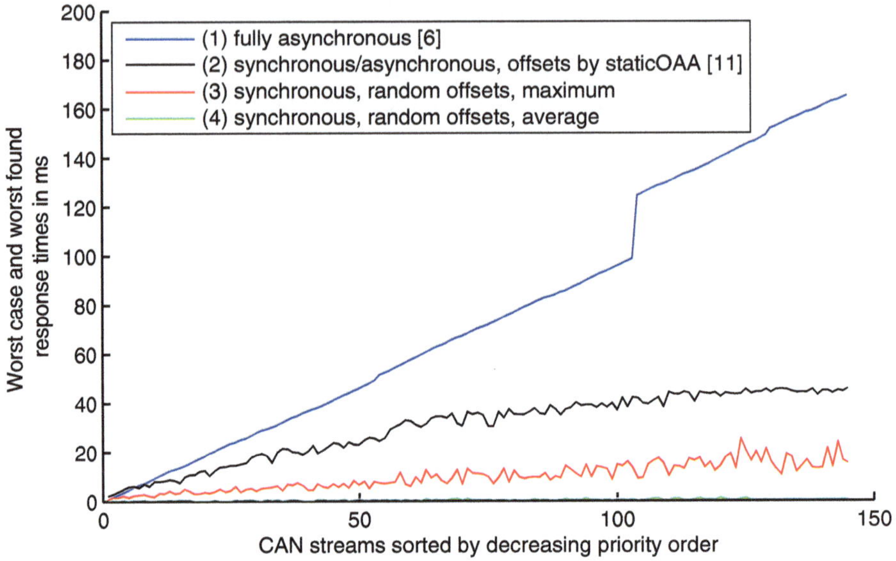

Fig. 5.2 Worst case response times and worst observed response times for different synchronism approaches for scenario 1

Comparison of different scheduling approaches typically uses WCRTs of all streams (or tasks). However, this procedure can in some cases not clearly distinguish whether one schedule is better than the other. For example one schedule tends to give advantage certain group of streams (e.g. those with low priority) but at the same time being worse for the streams with high priority. In addition, when the response times change over time, it would be good to have a single criterion equally suitable for all types of streams. In this work, we use the rating function called the average weighted worst case response time, or AWW, originally proposed in [25], which enables us to differentiate the quality of different message scheduling schemes:

$$AWW(t_a,t_b) = \frac{\sum_{i=1}^{k} \frac{WCRT_i(t_a,t_b)}{T_i}}{k} \tag{5.1}$$

where $WCRT_i(t_a,t_b)$ is the measured worst case response time of the stream i that has period T_i in the observed time interval between t_a and t_b and k is the total number of streams generating messages.

The function $AWW(t_a,t_b)$ takes into account that the streams with large periods are more sensitive to large response times, because each WCRT is weighted with its corresponding period. Finally, the sum is divided by the number of streams to get the average rating per stream. This also allows comparison of different scenarios with different numbers of streams.

It would be reasonable to think of additionally weighting the rating by the CAN priorities, but the priorities are already usually taken into account by the

Table 5.1 Rating function values for approaches from Fig. 5.2

Scenario	AWW
Fully asynchronous	0.18
Synchronous/asynchronous	0.084
Synchronous, maximum	0.030
Synchronous, average	0.003
Optimal	0

stream periods and, therefore, this would lead to overweighting of messages with low priorities. Furthermore, the WCRTs could be weighted by the designer if any additional knowledge of the importance of the streams is available. However, as this knowledge is usually very application-specific, in our experiments we decided to compare the schedules by using AWW. The rating function values AWW for the example in Fig. 5.2 are shown in Table 5.1, where a lower value of the rating function indicates a better system performance.

5.4 Dynamic Offset Adaptation

In our approach, we propose the dynamic adaptation of offsets, which change over time following the change of the traffic on the CAN bus. Because the system is distributed with no single point of system status and operation knowledge, the decisions to change offsets are based on traffic monitoring carried out by the individual nodes, independently each from the other. The two main differences between our proposed approach and the staticOAA [12] are that (1) we consider the entire distributed system, where offset adaptations are carried out on all nodes dynamically and independently, and (2) the offset adaptation on each node is initiated periodically and offset changed only if the conditions of the traffic require the change. In the staticOAA approach all offsets are calculated in advance and do not change during system lifetime.

We further extend our approach to the whole system lifetime, i.e., we include the initialization phase which the system passes after its start and before entering normal operation state, unlike the other approaches, which do not consider initialization at all. Once the system completes initialization it goes into normal operation state, where the adaptation of the offsets occurs dynamically. The adaptation of offsets is carried out automatically for each stream depending on the current traffic on the CAN bus by using the dynamic offset adaptation algorithm (DynOAA). The details of the algorithm as well as an example that illustrate its operation are presented in Sect. 5.4.1. Section 5.4.2 shows the solution to trapping of DynOAA in synchronous simulations. During the initialization phase, offsets are introduced by starting the first message of each stream in a predefined order as described in Sect. 5.4.3.

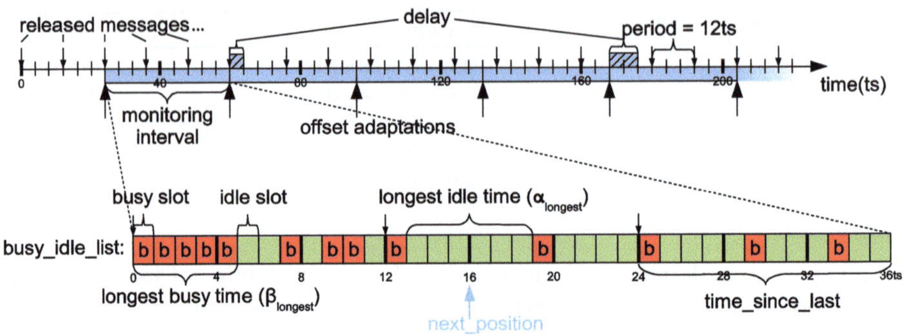

Fig. 5.3 DynOAA illustration—timing diagram and busy_idle_list on a single node

5.4.1 Dynamic Offset Adaptation Algorithm (DynOAA)

The DynOAA [26] is run on each node independently and periodically. An illustration of the operation of the DynOAA for one stream is shown in Fig. 5.3. In the upper part of the figure, on the top of the time line, the periodically released messages of the stream are indicated by small arrows. The larger arrows on the bottom of the time line indicate the instances when the adaptations start or when DynOAA is executed. Each algorithm run consists of a traffic monitoring phase and a delay phase. During the monitoring phase, a list called busy_idle_list is created. It is a circular list, meaning the last element is adjacent to the first element. An example of it is shown in the lower part of Fig. 5.3. It contains, for each time slot during the monitoring phase, an idle element if the bus is idle and a busy element if the bus is busy. If not mentioned otherwise a time slot has the length of the transmission time of one CAN bit. From the busy_idle_list, we can find the longest idle time $\alpha_{longest}$ and longest busy time $\beta_{longest}$, which are the maximum continuous intervals when the bus was idle or busy, respectively. During the delay phase, the next message of the stream is then delayed (i.e., the offset is adjusted) so that a message in the next monitoring phase is released in the middle of $\alpha_{longest}$ (next_position).

In distributed systems, all streams are considered independent of each other. If more than one stream starts to execute the adaptation simultaneously, there is a high probability that the value of the next_position will be identical at more than one stream. Instead of spreading, the message release times would in that case be clustered around the same time instant. Therefore, we need to ensure that only one stream is adapting its offsets at the same time. This is accomplished by introducing a unique criterion for all streams, which ensures that the stream that is the first in the longest busy time $\beta_{longest}$ will adapt its offset. All communication controllers have to adhere to this unique protocol individually.

The algorithm is best explained by the example from Fig. 5.3. A stream with a period of 12 time slots (ts) is considered. The transmission of one message is assumed to take one ts. The algorithm passes through two phases. In the first phase, monitoring, the busy-idle list in the lower part of Fig. 5.3 is created. The first message of the $\beta_{longest}$ belongs to the stream under consideration. Therefore, in the second phase, it will adapt, i.e., the next message release by this stream will be delayed. In order to calculate the needed delay, the next position is determined first by choosing the middle of the longest idle time, which is in this case 16 ts. The delay is then calculated as

$$\begin{aligned} delay &= (next\_position + time\_since\_last) \\ &\quad mod\, period \\ &= (16\,ts + 12\,ts)\, mod\, 12\,ts \\ &= 4\,ts \end{aligned} \tag{5.2}$$

This delay causes the releases of this stream in the next hyper-period to be shifted by additional 4 ts. For the messages at positions 1–4 this means, if they were delayed, their response time is reduced. The probability that a message will be delayed depends on the number of preceding messages. If the message is not preceded by any other message, the message cannot be delayed. Therefore, it makes sense to choose the first message from the $\beta_{longest}$ to adapt.

5.4.2 Adaptation Memory

A disadvantage of the proposed adaptation criterion is that the adaptation can oscillate between the same positions in certain situations, when the schedule cannot be improved further because the same offsets are repeatedly tried. An example is shown in Fig. 5.4, where stream 1 is scheduled back and forth between schedules 1 and 2. These oscillations can also happen over several adaptation steps. However, it can easily be avoided if each stream records its last m chosen delays in the adaptation memory, effectively meaning that each stream has to use m memory locations to store these delays. In each subsequent adaptation step, instead of taking one of the previously tried delays, the delay is decreased by one message length. The resulting behavior will be that the stream, in most cases, is not at the beginning of a longest busy period, and another stream will be chosen for adaptation. In schedule 3 in Fig. 5.4, the alternative next_position is shown.

We did several experiments with different sizes of adaptation memory. For example, Fig. 5.5 shows the rating of the current hyper-period over time using DynOAA with different m starting with random offsets. If no oscillation occurs, the schedule improves to a rating of zero where all message response times are zero. Otherwise, the rating value stagnates when the oscillation happens. From the experiment, we can conclude that when larger memory is used, the rating of the schedule decreases further and therefore, the probability of oscillations decreases. However, the relationship between the increment of memory increase

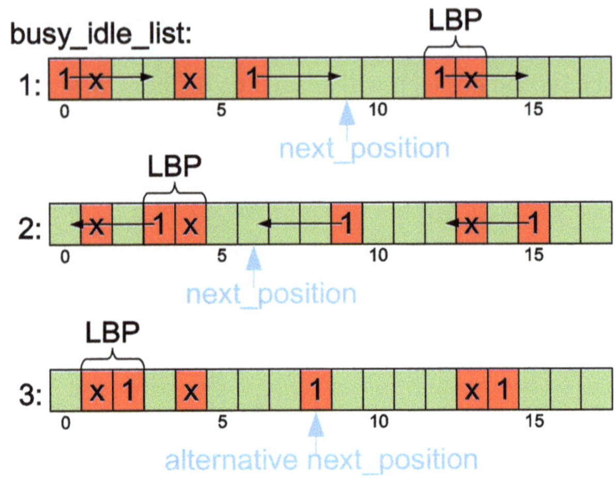

Fig. 5.4 Example where the adaptation gets stuck because it is repeatedly trying the same schedules. 1: Initial busy_idle_list, 2: regular adaptation, 3: alternative adaptation from initial state

and probability of oscillations is not linear. Further investigation of this phenomenon is left for future work, because the described problem only occurs for synchronous systems with periodic streams, like the one in our simulation. A real practical system has random components, such as non-periodic streams or different clocks. They will cause to leave the oscillating behavior and continue with the adaptation. Therefore, for all the following experiments we used an adaptation memory of the maximum integer value our simulation engine supported ($2^{31} - 1$).

How often the algorithm reaches an oscillating state and the time until leaving that state is difficult to predict. We have found in the experiments that using a small memory with ten entries is already sufficient to avoid oscillations for our test scenarios.

5.4.3 *Initialization*

A typical CAN bus system begins its operation by turning on all nodes simultaneously. This procedure is most often very simple to implement, because one shared source of power is used. From scheduling perspective, this synchronous startup would be desirable, as we could plan the initial offsets a priori.

However, in practical implementations the boot and initialization times of the different ECUs are not equal. They even vary from one start to another, because initialization programs take different time depending on different factors including the conditions of the physical environment. For example, if it is very cold, the rain sensor needs longer to provide initial measurements compared to when it is warm. Therefore, we have to assume random initial offsets. However, if the initial offsets are set randomly, then the worst case for an offset-free system becomes possible when all streams transmit at the same time.

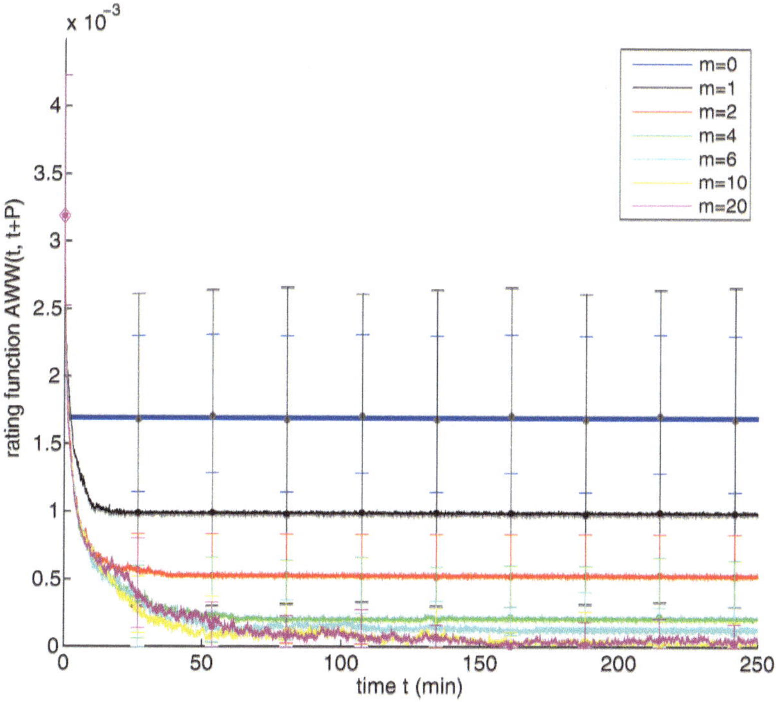

Fig. 5.5 Average of AWW of ten runs with different initial random offsets as a function of time for different sizes of memory for scenario 1. The error bars denote the minimum and maximum values.The legends are always in the same top-down order as the graphs

One possible way to alleviate this would be to measure the startup times of each ECU under different environmental conditions. Using these measurements, we could assign offsets a priory. However, in this case it is not feasible to apply the methods for single processor scheduling [13], because the offsets will not be fixed but represented with a range of values. Hence, a new method would be needed. In addition, the measurements will be very time consuming and could only be done very late in the system development phase. Also, any modification and revision in system design would require all offsets to be recalculated and changed on all ECUs. Therefore, we do not adopt this approach, but propose an

rithm that initializes the streams automatically not requiring any a priori knowledge.

The goal of the algorithm is to avoid the worst case in which all streams release messages at the same time, and to improve over a random initialization. The basic idea is that the streams start sequentially, one after another, in a predefined fixed order, in our case defined by stream priorities. They can start in either decreasing (highest priority first) or increasing (lowest priority first) priorities. This can be implemented in a very light-weight way, where each stream waits with the

Algorithm 1 Dynamic initialization algorithm

```
minDist = 1
waitTime = 1
priority_level = 0
min_dist_count = 0
wait_time_count = 0
for (each time slot) do
    if priority_level = priority on the bus then
        priority_level++
    end if
    if current slot is idle then
        if min_dist_count < minDist then
            min_dist_count++
        else if wait_time_count < waitTime then
            wait_time_count++
        else
            priority_level++
            wait_time_count=0
        end if
    else
        min_dist_count = 0
    end if
    if min_dist_count = minDist AND priority_level is this streams priority then
        start transmission
    end if
end for
```

transmission of its first message until the stream with next higher (or lower) priority successfully transmits a first message. The problem can be that not all priorities are assigned to the streams, effectively meaning that a stream may be indefinitely waiting on transmission of a message with immediately higher (or lower) priority. To avoid this situation, all streams keep track of the priority of the last released stream. If the next expected priority is not assigned to any node, the stream with the next lower (higher) priority will start. The unused priority value can be detected because all streams know when the first message should be sent. By simply waiting for the transmission of all priority IDs for a given amount of time (using a timeout), unused priorities can be identified and simply neglected.

In Algorithm 1 a detailed description of the initialization procedure is provided. The two parameters *minDist* and *waitTime* need to be chosen in advance. MinDist is the number of idle time slots between two scheduled messages. A time slot is defined similar to a time slot from Sect. 5.4.1. WaitTime is the number of idle time slots the algorithm has to wait until it can be assumed that the expected priority level is not sending and can be skipped. The algorithm counts idle time slots until minDist is reached. If the expected priority level equals to the streams priority, it starts sending. Otherwise, the following idle time slots are counted until waitTime is reached. For every time reaching waitTime, the priority level is increased. The

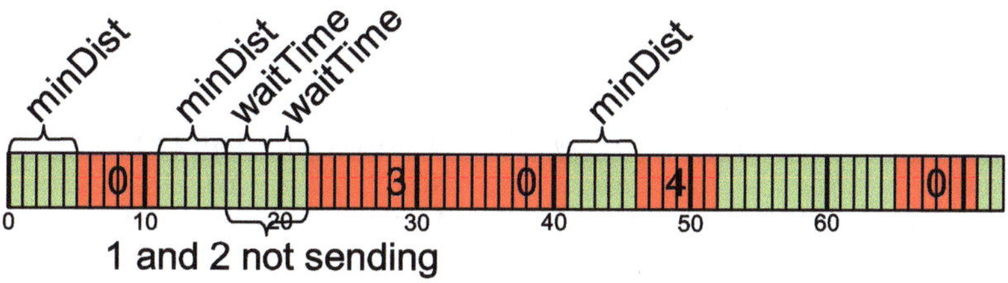

Fig. 5.6 Example schedule of the dynamic initialization algorithm with minDist = 5 and wait Time = 3. The stream with priority zero has a period of 30 time slots. The streams with priority one and two are not present in the system and are therefore skipped after 11 idle time slots

example in Fig. 5.6 clarifies the function of the algorithm and the meaning of the two constants.

If an ideal system is assumed, both constants, minDist and waitTime can be set to one. In an ideal system as represented by our simulations the nodes have zero computation time and are perfectly synchronous. If not mentioned otherwise, this setting is used for the experiments in this paper. In a real system the constants need to be increased. A more detailed discussion and reasoning is found in Sect. 5.5.2.3, where experimental results are presented. In conclusion, with the proposed approach the system can be fully automatically initialized. In contrast to a random initialization approach, the response times from the start are reduced. The algorithm does not depend on a simultaneous start of all streams, because in the worst case the stream is scheduled after one hyper period. This allows flexible activation and deactivation of streams.

5.5 Results and Comparisons

In this section, we compare DynOAA with previous works. We first make a qualitative comparison of the selected important properties. For the quantitative comparison, different scheduling approaches are compared and analyzed for a number of scenarios from literature.

5.5.1 *Qualitative Comparisons*

In Table 5.2, different approaches for scheduling and analyzing CAN are compared against the following properties:

Flexibility: It takes into account whether the approach can only cope with fixed scenarios, or it is flexible enough to handle addition or removal of streams during run-time.

Table 5.2 Qualitative comparison of different CAN scheduling strategies

	Schedulability analysis [7]	Random offset assignment	Staticoaa [12]	DynOAA	FTT-CAN[2]
Flexibility	No	Yes	No	Yes	Yes
Response times	Bad	Good	Average	Very good	Very good
Communication overhead	No	No	No	No	Yes
Computation overhead	Offline low	No	Offline low	Online low	Online high
Predictability	Yes	No	Yes	Partial	Yes

Response times: The response times of messages are very important because if they are too long the system might fail.

Computation/communication overhead: This describes how often and how much additional computation/communication the approach requires.

Predictability: It shows whether it is possible to calculate bounds for the actual response times.

The schedulabilty analysis and random offset assignment approaches use the basic CAN access mechanism allowing random access of the streams. Using schedulability analysis, however, pessimistic worst case response times are calculated and used as upper bounds. The StaticOAA approach goes a step further and assigns static offsets between the streams on one ECU. The clock on the ECU guarantees that these offsets remain fixed and analysis can be used to calculate worst case response times. The DynOAA approach extends the assignment of offsets to the whole distributed system, adapting the offsets dynamically. A different strategy is to set a protocol on top of CAN. FTT-CAN [2] is in our opinion the best option, because it provides the flexibility of scheduling both event-triggered and time-triggered message streams.

For the first three properties, DynOAA is better or equal to the previous works. DynOAA has full flexibility similar to random offset assignment. FTT-CAN is also able to add and remove streams at run-time, but it incurs communication overhead due to registration at the master node. The message response times of DynOAA are shorter compared to the first three strategies using the AWW as showed in the following section. Because of the mixed time- and event-triggered scheduling of FTT-CAN, it is difficult to compare the response times with DynOAA. The event-triggered streams have response-times similar to the random offset assignment, while the time-triggered streams have zero response times if the tasks are executed according to the FTT-CAN schedule. The communication overhead of DynOAA cannot be further reduced as there is none. The two properties we have to take a closer look at are: computation overhead and predictability. The schedulability analysis and staticOAA offsets are determined before system deployment and actually do not introduce any run-time overhead. DynOAA and FTT-CAN are based on run-time calculations. Using FTT-CAN, a single master node calculates the schedule for the whole network, while DynOAA is executed on all nodes calculating only the local behavior that reduces the amount of computation needed drastically. The predictability when using FTT-CAN is better, because it supports time-triggered access. Schedulability analysis and staticOAA are based on analytical models and, therefore, can provide guaranteed WCRTs while our approach lacks analytical analysis and guarantee. However, our approach is not as unpredictable as a purely random approach, as we demonstrate by experiments that show the schedule created by DynOAA always improves over time and the actual response times will never reach the values calculated by the analytical approaches.

5.5.2 *Quantitative Comparisons*

In the following section, the performance of our approach is quantified by means of simulation using typical automotive scenarios.

5.5.2.1 Simulation

We developed and used our own CAN bus simulator, because other known available simulators, such as for example RTaW-Sim [20], are not able to deal with all scenarios we apply and cannot extract the required figures for the properties explored. Our simulator is discrete-event driven with the simulation step equal to one CAN bit. We follow an object-oriented approach using Java as the design language. The main flow of the simulation is depicted in Fig. 5.7. The flow consist of three steps: Loading of the scenario, simulating each bit for a fixed number of time steps and then storing the gathered results in external files.

The first step is to initialize the simulation by loading the scenario. The scenario is stored in an xml-file according to the specification of Netcarbench [4]. For each message stream it contains a name, a priority, a period, a length, an offset and a deadline. Most of the simulation time is spent in the second step. Therefore, we will detail the functional principle of this step. Three different objects are mainly responsible for the simulation: Sim, Bus and Stream. The Sim object contains the main method and controls the course of action. The Bus object stores the information of the current bus state, while the Stream object stores the information only for its stream. Two example operating sequences are shown in Fig. 5.8. In the first sequence, the bus is free, which is indicated by the Bus. At the beginning, the Sim asks all streams whether they have something to send. Next, the streams that have something to send are notified whether they are sent or not. The decision which one will be sent is based on the priority of the streams. Finally, the Bus and all Streams are informed which Stream sent a message. The Bus uses this information

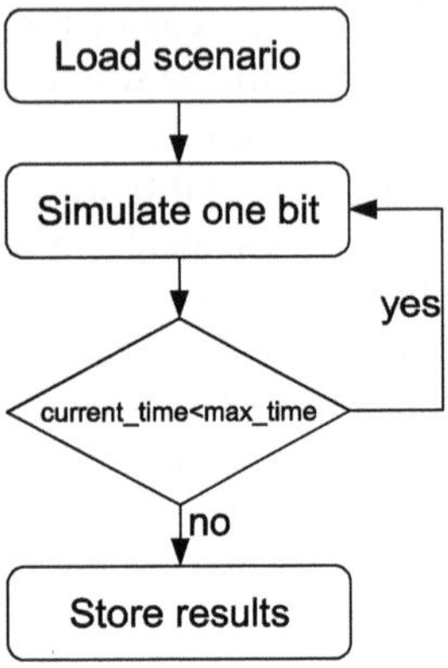

Fig. 5.7 Main flow of the simulation

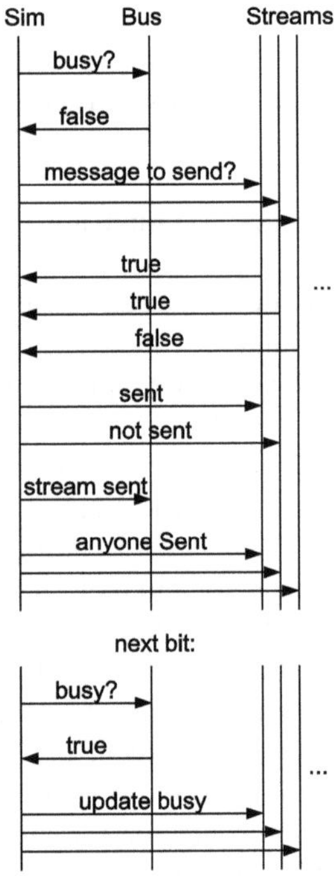

Fig. 5.8 Two example operating sequences of the main simulation engine

to calculate how long the message will occupy the bus. The length of each message is calculated by assuming worst case bit-stuffing. The second sequence shows the case where the bus is busy.

Because the Streams contain the information for generating the AWW and because our adaptation algorithm is integrated into the Stream's behavior, this object

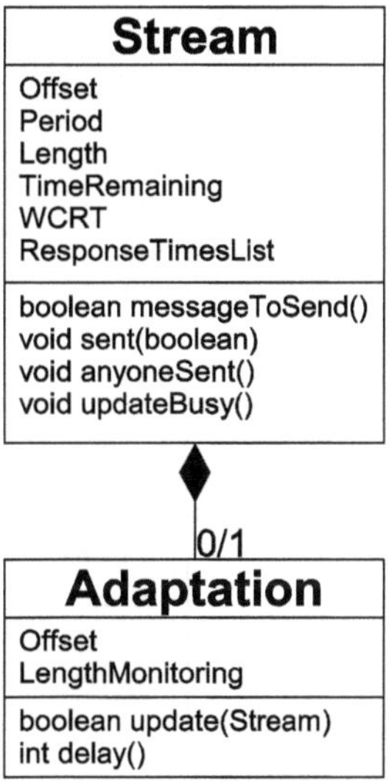

Fig. 5.9 Class diagram of the Stream class

is described in more detail. Figure 5.9 shows the class diagram of the Stream. Besides the static parameter of a message stream, the Stream stores the number of time steps until the next message should be transmitted, the worst case response time, and a list of the last response times. If the stream is adapting the offsets, it has an Adaption object. It needs to be updated every bit and the calculated delay needs to be requested when the update function returns true, indicating that the end of a monitoring interval has been reached.

A consequence of the discrete event simulation is that only the synchronous case, where all nodes have the same time base, can be simulated. This means, if the offsets are fixed and not adapted by DynOAA, the schedule repeats after time equal to the hyper-period. The asynchronous case is then modeled and simulated by using different randomly selected initial offsets. The simulation assumes the error free case.

5.5.2.2 Scenarios

For our experiments we used scenarios common in the automotive domain as described in Table 5.3. The scenarios are split into three groups: (1) Two scenarios provided with the RTaW-Sim [20] (ls_50, hs_50a), (2) an example provided by Herpel et al [15] (hs_50b), and (3) four synthetic scenarios generated by Netcarbench [4](hs_60 to hs_90). Netcarbench allows generating typical automotive scenarios for a given workload. Workload in this context is $\sum_{i=1}^{k} \frac{C_i}{T_i}$. We generated scenarios with a bus speed of 500 kbits with different average loads. The nomenclature for the scenarios is the following. Scenarios with a CAN bus speed of 125 kbits start with

Table 5.3 Scenarios used for the experiments

Scenario	Speed (kbits)	No of streams	Workload	Max period	Source
ls_50	125	145	0.51	2s	[20]
hs_50a	500	85	0.51	2s	[20]
hs_50b	500	56	0.48	1s	[15]
hs_60	500	97	0.60	2s	[4]
hs_70	500	101	0.71	2s	[4]
hs_80	500	109	0.81	2s	[4]
hs_90	500	131	0.90	2s	[4]

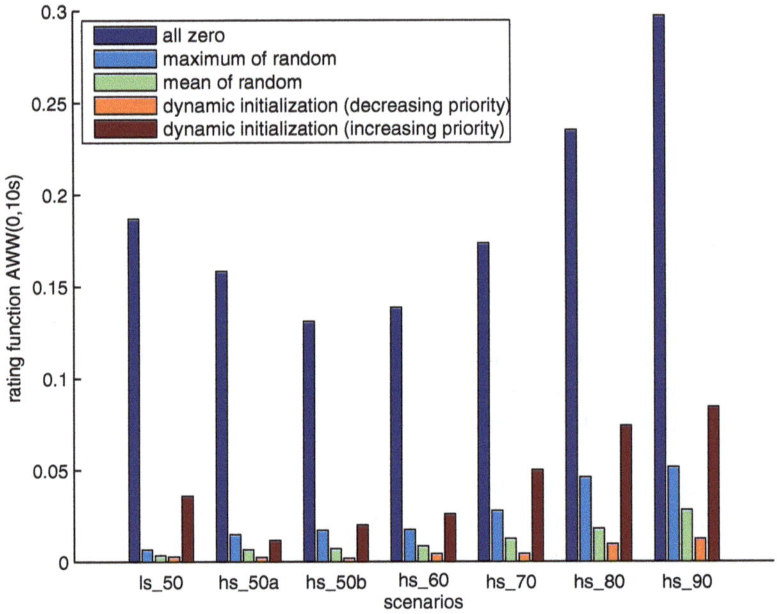

Fig. 5.10 AWW for different initialization strategies

ls (low-speed) and with 500 kbits start with *hs* (high-speed). The number at the end describes the average bus load of the scenario. For example, hs_60 is a scenario with the bus speed of 500 kbits with a workload of 60 %. All scenarios are compared in terms of AWW defined in Sect. 5.3.2.

5.5.2.3 Effects of Initialization

First, we compare different initialization phase strategies as shown in Fig. 5.10. The worst case is when all offsets are zero and all streams start to transmit simultaneously.

Although this is an overly unlikely scenario, it serves as a reference and indication of the upper bound. The second strategy is to initialize the offsets randomly. For this case, we made 1,000 simulation runs with uniformly distributed offsets and show the maximum and mean value of these simulation runs. The last two cases represent our two proposed dynamic initialization strategies, one with decreasing and the other with increasing priorities. In this simulation, we assume all streams are active from the start.

Figure 5.10 shows that the dynamic initialization with decreasing priorities is always the best strategy. The dynamic initialization with increasing priorities performs worse than the maximum of the random initialization except for scenario hs_50a. This is caused by the fact that the priorities of the streams are mainly assigned according to their periods. If the streams with smaller periods are scheduled later, then they have releases at the beginning of the next hyper period. These releases can conflict with the earlier scheduled messages of the streams with larger periods. This is not the case if the streams are scheduled before their period, which happens more often when starting with the lowest priorities.

Next, we analyze the effects of the two constant parameters waitTime and minDist. A real system has inaccuracy that could lead to malfunction of the algorithm. For example, a stream could miss its slot because it didn't respond fast enough. The parameters need to be increased accordingly to avoid these problems. At the current state of development it is impossible to measure the effects on a real system, because the inaccuracy is highly dependent on the underlying hardware. However, we can estimate the effect by simulation with increased parameters.

Figure 5.11 presents the results of simulating the dynamic initialization for the scenario ls_50 and hs_90. For scenario ls_50, it shows that with increased waitTime the rating increases, but it is always better than the average of random initialization. Also, the parameter minDist has no influence. For the fully loaded scenario hs_90, the parameter minDist has some influence on the rating. However, only when both parameters are above 50, the rating is in the range of the mean random initialization.

Although the results of simulation of the initialization show the approach is robust in terms of variation of parameters, further research on a real system prototypes needs to be done to confirm that the approach will work with the same results in real systems.

5.5.2.4 Rating Over Time

Figure 5.12 shows the AWW that uses the maximal response times recorded since the start of the simulation. These values are comparable to the WCRTs of the analytical analysis [7,14], as they reflect the worst response time that is found during our simulation. The AWW for the different approaches are listed in Table 5.4. Even though we cannot prove that the WCRTs have reached their maximal value, the simulation experiments show that the rating function is increasing very slowly over time for a reasonable amount of simulated time (1,400 min $\approx$ 1 day).

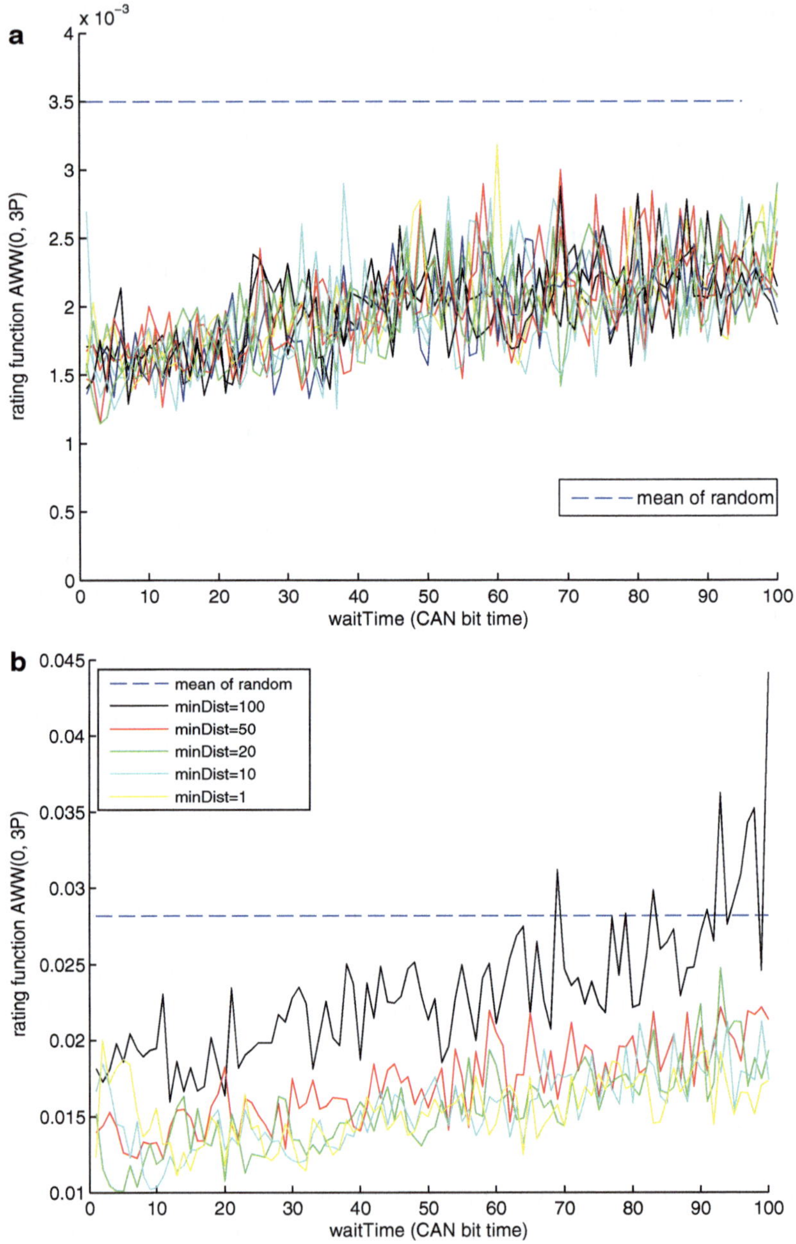

Fig. 5.11 Rating of the schedule after initializing all streams for different parameter settings. Both parameters are chosen in multiples of the time for one CAN bit (8us for ls_50, 2us for hs_90)

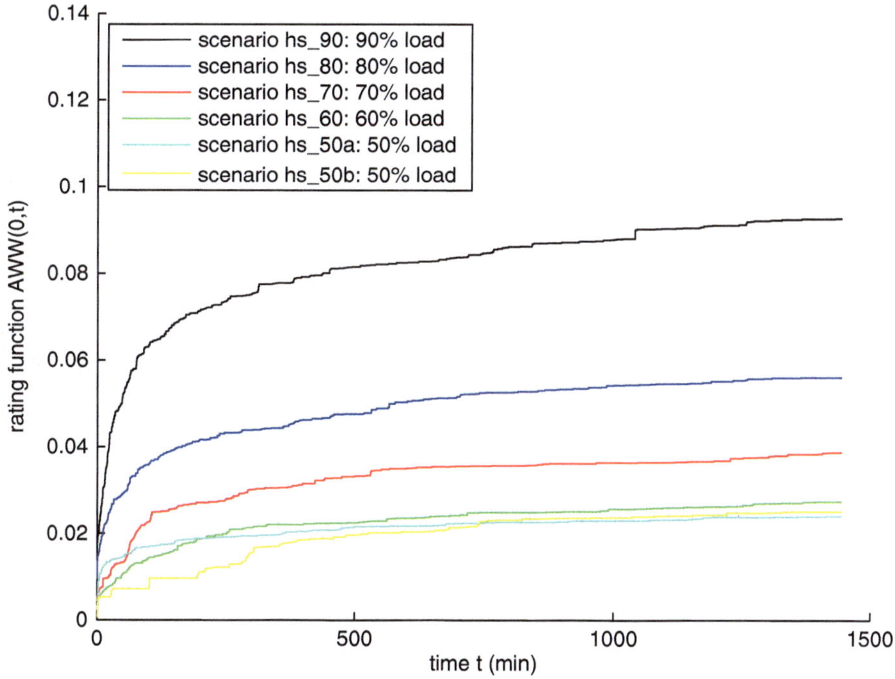

Fig. 5.12 Rating function AWW as the function of time—maximal response times for the entire simulation taken into account

Table 5.4 AWW for different scenarios, where the first and second column take into account the WCRTs of one hyper period, the third column of the entire simulation time and the last column shows the average AWW for the last hyper period

Scenario	Analytical [7] $(\cdot 10^{-1})$	StaticOAA [14] $(\cdot 10^{-1})$	DynOAA (entire sim) $(\cdot 10^{-1})$	DynOAA (average) $(\cdot 10^{-3})$
ls_50	11.9	10.79	10.07	01
hs_50a	11.6	10.72	10.20	01
hs_50b	11.3	ln/a	10.21	01
hs_60	11.4	10.63	10.26	1.81
hs_70	11.7	11.03	10.40	4.01
hs_80	12.4	11.10	10.59	6.61
hs_90	13.0	12.84	11.0	121

Figure 5.13 shows the rating function calculated only for the last adaptation interval (here 1s). We can see that it converges very fast to a low average value. If we are dealing with a soft real-time system, where a few long response times can be tolerated, our method offers excellent performance.

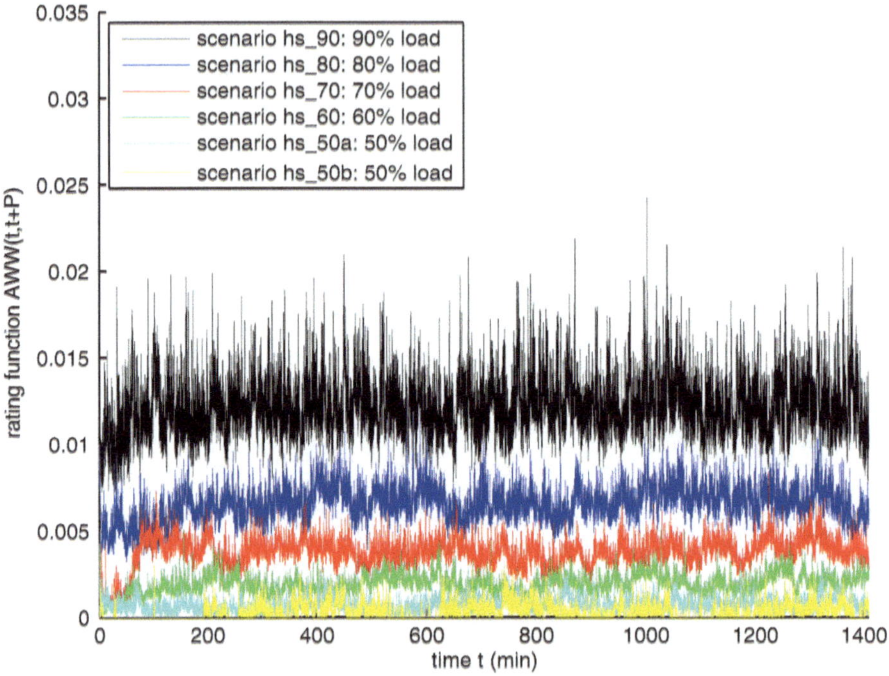

Fig. 5.13 Rating function AWW as the function of time for different application scenarios—response times of the last adaptation interval taken into account

5.6 Conclusions and Future Work

In this paper, we propose a new approach to scheduling messages on CAN that results in its better utilization and the reduction of average message response times compared to other known approaches under high loads. The assumption of asynchronous nodes/streams on CAN in our approach is more realistic than the assumptions used in other known approaches. Our approach can be fully automated by dividing system operation into initialization and normal operation phase. During the normal operation phase, message scheduling is performed by using the dynamic offset adaptation algorithm (DynOAA) on each system node. As the result, it allows designers to reduce the level of pessimism typically applied in scheduling messages on CAN. The reduced, and sometimes optimal, response times are achieved dynamically and without the need for specific prior or continuously acquired knowledge at any single point in the system, but rather in fully distributed and autonomously performed algorithm executed on all nodes of the system. As the algorithm does not require any knowledge of the overall system configuration and number of nodes, it automatically adapts to the changes of the system configuration (e.g. plugging in new or disconnecting the existing nodes). Comparisons with other approaches have been done by using a new performance measure in the form of rating function that calculates the averaged weighted worst case response times of individual streams. The results have been validated by extensive simulations for which a new simulator has been developed.

Our future work is focused along multiple paths. We plan to explore other rating functions to measure and compare the quality of different approaches for message scheduling, as well as applicability of the approach to other types of networks, including some wireless alternatives. Finally, our immediate plan is to work on the implementation of the algorithm and verification of its properties on the nodes of real CAN based systems.

Acknowledgements This work was supported in part by the German Research Foundation (DFG) under contract TE 163/15-1.

References

1. Albert A (2004) Comparison of event-triggered and time-triggered concepts with regard to distributed control systems. Embedded World 2004:235–252
2. Almeida L, Pedreiras P, Fonseca J (2002) The FTT-CAN protocol: why and how. IEEE Trans Ind Electron 49(6):1189–1201. DOI 10.1109/TIE.2002.804967
3. CAN in Automation, C.: Canopen. http://www.canopen.org/
4. Braun C, Havet L, Navet N (2007) NETCARBENCH: A benchmark for techniques and tools used in the design of automotive communication systems. In: IFAC Secretaria 7th IFAC international conference on fieldbuses and networks in industrial and embedded systems, pp 321–328
5. Bueno E, Hernandez A, Rodriguez F, Giron C, Mateos R, Cobreces S (2009) A dsp- and fpga-based industrial control with high-speed communication interfaces for grid converters applied to distributed power generation systems. IEEE Trans Ind Electron 56(3):654–669
6. CAN specification 2.0 b (1991) Robert Bosch GmbH, Stuttgart, Germany
7. Davis R, Burns A, Bril R, Lukkien J (2007) Controller area network (CAN) schedulability analysis: Refuted, revisited and revised. Real-Time Syst 35(3):239–272
8. Diaz J, Rodriguez E, Hurtado L, Cacique H, Vazquez N, Ramirez A (2008) Can bus embedded system for lighting network applications. In: IEEE 51st midwest symposium on circuits and systems, 2008. MWSCAS 2008, pp 531–534. DOI 10.1109/MWSCAS.2008.4616853
9. Felser M (2005) Real-time ethernet – industry prospective. Proc IEEE 93(6):1118–1129. DOI 10.1109/JPROC.2005.849720
10. FlexRay Consortium (2009) FlexRay communications systems - protocol specification. http://www.flexray.com. Accessed on 9 October 2012
11. Gamiz J, Samitier J, Fuertes J, Rubies O (2003) Practical evaluation of messages latencies in CAN. In: IEEE conference on emerging technologies and factory automation, 2003. Proceedings of ETFA'03, pp 185–192
12. Gil-Castineira F, Gonzalez-Castano F, Franck L (2008) Extending vehicular can fieldbuses with delay-tolerant networks. IEEE Trans Ind Electron 55(9):3307–3314. DOI 10.1109/TIE.2008.927972
13. Goossens J (2003) Scheduling of offset free systems. Real-Time Syst 24(2):239–258
14. Grenier M, Havet L, Navet N (2008) Pushing the limits of CAN-scheduling frames with offsets provides a major performance boost. In: Proceedings of the 4th European congress embedded real time software (ERTS 2008), Toulouse, France
15. Herpel T, Hielscher K, Klehmet U, German R (2009) Stochastic and deterministic performance evaluation of automotive CAN communication. Comput Networks 53(8):1171–1185

16. Marino P, Poza F, Dominguez M, Otero S (2009) Electronics in automotive engineering: A top-down approach for implementing industrial fieldbus technologies in city buses and coaches. IEEE Trans Ind Electron 56(2):589–600. DOI 10.1109/TIE.2008.2002723
17. Navet N, Simonot-Lion F (2009) Automotive embedded systems handbook. CRC, West Palm Beach
18. Nolte T, Nolin M, Hansson H (2005) Real-time server-based communication with can. IEEE Trans Ind Inform 1(3):192–201. DOI 10.1109/TII.2005.852074
19. Prodanov W, Valle M, Buzas R (2009) A controller area network bus transceiver behavioral model for network design and simulation. IEEE Trans Ind Electron 56(9):3762–3771. DOI 10.1109/TIE.2009.2025298
20. RTaW-Sim (2009) Real-time at Work CAN Simulator. http://www.realtimeatwork.com/software/rtaw-sim/. Accessed on 9 October 2012
21. Samii S, Rafiliu S, Eles P, Peng Z (2008) A simulation methodology for worst-case response time estimation of distributed real-time systems. In: Proceedings of the conference on design, automation and test in Europe. ACM, New York, pp 556–561
22. Sommer J, Burgstahler L, Feil V (2006) An analysis of automotive multi-domain can systems. Proceedings of the 12th EUNICE Open European summer school. Accessed on 9 October 2012
23. Stankovic J, Ramamritham K, Spuri M, Buttazzo G (1998) Deadline scheduling for real-time systems. Springer, Berlin
24. Zhou F, Li S, Hou X (2008) Development method of simulation and test system for vehicle body CAN bus based on CANoe. In: 7th world congress on intelligent control and automation, WCICA, pp 7515–7519. IEEE, New York
25. Ziermann T, Teich J (2010) Adaptive traffic scheduling techniques for mixed real-time and streaming applications on reconfigurable hardware. In: IEEE Proceedings of 17th reconfigurable architectures workshop (RAW)
26. Ziermann T, Salcic Z, Teich J (2011) DynOAA – dynamic offset adaptation algorithm for improving response times of CAN systems. In: IEEE Proceedings of design, automation, and test in Europe (DATE), pp 269–272

Chapter 6
Self-Configurable Automotive Embedded Systems

Achim Rettberg and Ina Podolski

Abstract In this paper a middleware architecture for distributed automotive systems that supports self-configuration by dynamic load balancing of tasks is presented. The inclusion of self-configurability is able to offer reliability within the multimedia network of the vehicle (Infotainment). Load balancing of tasks could be applied if an error occurred within the network. The error detection in the network and the load balancing should run automatically. Therefore, the middleware architecture has to deal on one hand with the error detection and on the other hand with the migration of tasks. Additionally, to enable the migration it is important to identify the requirements of all electronic control units (ECU) and tasks within the network.

6.1 Introduction

Future application scenarios for vehicle electronic systems include on one side the access to mobile devices that build ad-hoc networks with the built-in devices of the vehicle and on the other side the support of robustness, redundancy and dependability within the vehicle network. Modern electronic vehicle networks have to be as flexible as possible to cope the actual requirements. The idea to build up a self-configurable system could help to overcome these requirements. A mobile device could automatically be attached and integrated in the existing system if the system supports self-configurability. If an ECU has a failure all task could be mi-grated to other ECUs inside the vehicle network by a self-configurable system middleware. Self-configuration could be applied to distribute networks. In modern vehicles three types of networks are built in. That is namely the Powertrain, Body

A. Rettberg (✉) • I. Podolski
Carl von Ossietzky University Oldenburg, Germany
e-mail: achim.rettberg@iess.org; ina.podolski@iess.org

M.T. Higuera-Toledano et al. (eds.), *Self-Organization in Embedded Real-Time Systems*, DOI 10.1007/978-1-4614-1969-3_6,

or Chasis and Infotainment network. Within the Powertrain safety critical tasks like the anti-blocking system and the motor management are located.

The Body or Chasis network contains also critical tasks, but the vehicle will run if a failure occurs. The window opener is a typical task of this network. The Infotainment network consists of more or less media based tasks, like the radio or navigation system. Due to safety critical reasons our approach will focus on the Infotainment network.

To increase the quality of the vehicle it is important to built in fault-tolerant systems in the network. In a distributed system fault-tolerance can be include in three ways: Replication, redundancy, and diversity. While the former provides multiple identical instances of a system, the tasks and requests are directed to all of them in parallel, and the choosing of the correct result is based on a quorum, redundancy is characterized by multiple identical instances and a switching to one of the remaining instances in case of failure. Diversity provides different implementations of a system that are used like replicated systems. A self-configurable system is able to provide redundancy, diversity and replication of tasks, therefore, it helps to make the system more stable.

In the context of self-configuration of automotive systems redundancy of data, applications, and tasks can be used to get an increased fault-tolerance in case of ECU failures. Crucial data, applications or tasks are distributed as backup components on the ECUs of the vehicle system that they can be used of or executed by other ECU if their originally ECUs failed. The way of distribution and the number of replicas and the decision which components are replicated depends on an adequate algorithm. At this costs of replication and migration and load of other ECUs will be considered.

In the following a vehicle middleware architecture is presented that supports self-configuration by load balancing strategies for non-critical tasks within the Infotainment network. In our case we enable a dynamic reconfigurable system by load balancing. In existing approaches self-configuration is enabled, by including redundancy and replication of tasks during design time. This is a static system re-configuration. Furthermore, our middleware offers services to realize a load balancing based on different strategies for the Infotainment network. This work is part of the DySCAS project [1]. The main objective of the DySCAS project is the elaboration of fundamental concepts and architectural guidelines, as well as methods and tools for the development of self-configurable systems in the context of embedded vehicle electronic systems. The reason is the increasing demand on configurational flexibility and scalability of the systems imposed by future applications which will include simultaneous access to a number of mobile devices and ad-hoc networking with the built-in devices.

The rest of the paper is organized as follows: Sect. 6.2 will describe the related work in the field of research where our architectural approach is located. As a motivation for this paper Sect. 6.3 motivates and describes a use case scenario. Afterwards our middleware architecture is presented, see Sect. 6.4. In Sect. 6.5 we describe the load balancing strategy we use within the middleware. A short description of the algorithm and simulation are discussed in Sect. 6.6. We conclude the paper with a summary and give an outlook for future work.

6.2 Related Work

In this section we will give a short overview of existing load balancing approaches to support self-configuration and on middleware approaches in automotive systems. There are several publications regarding load balancing and extensive research has been done on static and dynamic strategies and algorithms [9].

On the one hand, load balancing is a topic in the area of parallel and grid computing, where dynamic and static algorithms are used for optimization of the simultaneous task execution on multiple processors. Cybenko addresses the dynamic load balancing for distributed memory multiprocessors [4]. In [8] Hu et al. regard an optimal dynamic algorithm and Azar discusses on-line load balancing. Moreover Diekmann et al. differentiate between dynamic and static strategies for distributed memory machines [5].

Heiss and Schmitz [7] introduce the Particle Approach that deals with the problem of mapping tasks to processor nodes at run-time in multiprogrammed multicomputer systems solved by considering tasks as particles acted upon by forces. All these approaches have the goal of optimizing the load balancing in the area of parallel and grid computing by migrating tasks between different processors, while our approach focuses the direct migration of selected tasks to a newly added resource. Furthermore we regard load balancing that is located on the middleware-layer.

Moreover there are static approaches, like [19], that address a finite set of jobs, operations and machines, while our approach deals with a dynamic set of tasks and processors within the vehicle system. Balasubramanian, Schmidt, Dowdy, and Othman consider in [12, 16], and [15] middleware load balancing strategies and adaptive load balancing services. They introduce the Cygnus, an adaptive load balancing/monitoring service based on CORBA middleware standard. Their concept is primarily described on the basis of a single centralized server, while decentralized servers that collectively form a single logical Load Balancer is not explained in detail. Moreover the topic of dynamic reconfigurable automotive systems is regarded in [1, 2, 10, 11]. In the following paragraphs we discuss several middleware approaches for automotive systems.

The Autosar consortium (www.autosar.org) suggested a middleware approach based on a run-time environment (RTE). The RTE is developed to support a common infrastructure for automotive systems. The self-configurability developed in our approach will enrich the Autosar RTE especially by dynamic reconfiguration management through load balancing. In [17] a formal specification for developing distributed, embedded, real-time control systems is described. The middleware supports dependable, adaptive dynamic resource management based on replicated services.

An additional approach according fault-tolerance and dynamic reconfiguration is discussed in [3]. Again replicated services are used in this model. In [13] a middleware architecture for telematics software based on OSGi and AMI-C specification is presented. An application manager is introduced for telematic applications. The architecture enable in-vehicle terminal to provide various telematics services to increase driver's safety.

The authors of [14] describe trends for automotive systems. They give an over-view of requirements for middleware systems in this area. Especially what indus-try demands for such middleware services. Hiding the distribution and the heterogeneity of such platforms is demanded as well as providing high-level services (e.g. mode and redundancy management) and ensuring QoS.

6.3 Motivation

In this section we give a motivation for our approach. We identify three use cases:

- Task fails on an ECU and have to migrate to another one
- ECU has a defect—all task will be migrated
- New device is attached to the network

As an example we will use the second use case. If an ECU of the vehicle Infotainment system failed, a migration to another ECU within the vehicle that is able to execute the applications or tasks should be possible. Thus it is possible to mi-grate for example tasks of the ECU with the radio system to the ECU running the navigation system.

After the failure occurred within the vehicle the system starts a self-reconfiguration without avoiding overloading ECUs. The self-reconfiguration is surely based on specific characteristics from the tasks and the ECUs. That means, it has to be ensured that a task could only run on an ECU that is able to execute it.

In consideration of all running processes and the resources situation within the vehicle network appropriate services decide on a possible load balancing according to different strategies and initiate the task migration where required. Thus in our example where an error occurred inside the radio system the appropriate tasks migrate from the radio to the navigation system. Let us assume that the navigation system respectively the ECU is able to run the tasks from the radio system.

6.4 Proposed Middleware Architecture

To realize the use case scenario (failure in the radio system) described above and other possible services for example device detection a middleware architecture is required that fulfills several requirements. We introduce four sub-modules to handle self-configuration in the middleware. The Event Management detects failures in the vehicle network and it is responsible for detection and removal of additional ECUs. Detailed information and capabilities of existing ECUs as well as the registration of newly added devices is realized within the Device Registration module. All status information and the resource load of each ECU within the vehicle are stored by the Resource Management. Finally, the Load Balancing initiates the task migration based on specific characteristics and requirements of the tasks and ECUs. In the following we give a more detailed view of the middleware.

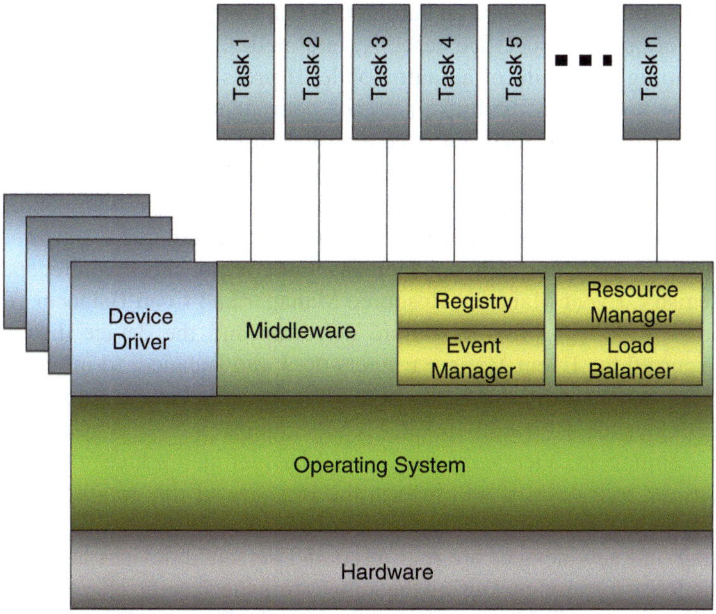

Fig. 6.1 Self-configurable architecture

The operating system builds the interface between the hardware and the middleware (see Fig. 6.1). Additionally, device drivers are necessary for specific hardware parts. The tasks run on top of the middleware. Middleware is a software layer that connects and manages application components running on distributed hosts. It exists between network operating systems and application components. The middleware hides and abstracts many of the complex details of distributed programming from application developers. Specifically, it deals with network communication, coordination, reliability, scalability, and heterogeneity. By virtue of middleware, application developers can be freed from these complexities and can focus on the application's own functional requirements.

Before explaining the design of our automotive middleware and the specific services, we enumerate the five requirements of automotive middleware. These requirements are resource management, fault-tolerance, and specialized communication model for automotive networks, global time base, and resource frugality. These requirements are derived from the distributed, real-time, and mission-critical nature of automotive systems and differentiate automotive middleware from conventional enterprise middleware products.

A vehicle has a real-time nature. It is a system in which its correctness depends not only on the correctness of the logical result, but also on the result delivery time. Since a vehicle is subject to various timing constraints, every component in a vehicle should be designed in a way that its timing constraints are guaranteed a-priori. At the same time, the timing constraints of a vehicle should be guaranteed in an end-to-end manner since an automobile is a distributed system and its timing constraints are usually specified across several nodes. For example, let us consider a typical timing constraint of an automobile. If pressing a brake pedal is detected at the sensor node, then the brake actuator node must respond to it within 1 ms. To meet this constraint, there must be a global Resource Manager that calculates the required amount of resources on each node and actually makes resource reservations to network interface controllers and operating systems on distributed nodes. Automotive middleware is responsible for such resource management.

The middleware in our approach includes four components that offer specific services: Registry, Event Manager, Resource Manager and Load Balancer. The Event Manager is responsible for the failure detection and the device discovery. If a failure occurred the Event Manger triggers the Load Balancer to initiate a feasible migration of tasks. Additionally, if a new device is added to the automotive system via technologies like Bluetooth or WLAN for example, it is recognized by the Event Manager component. Vice versa the Event Manager also notices the detaching of the device. In both cases it will inform the Registry of the middleware about the availability or the detaching of the additional device.

Existing and new devices are registered and detached devices are unsubscribed within the Registry service. During the registration the specific characteristics of the device (like memory, CPU, etc.) are stored within the Registry. Due to the distributed system the Registries of each vehicle ECU (Electronic Control Unit) communicate with each other to guarantee that each Registry of an ECU knows the actual status of all devices within the network inclusive of the newly added devices.

The Load Balancer spread tasks between the vehicle ECUs in order to get optimal resource utilization and decrease computing time. It evaluates possible migration of tasks based on different load balancing strategies. To guarantee a suitable migration the Load Balancer considers the current resource situation on the ECUs with aid of the Resource Manager. If a failure is occurred the Load Balancer tries to find based on the characteristics of the tasks and ECUs a feasible migration. Once a load balancing on an additional device is started, and this device is detached while the migrated tasks are executed, they will be re-started on the original ECU again. In this case the Event Manager is responsible to inform the Load Balancer to initiate this re-start.

The Resource Manager supervises the resources of the local ECU. To be aware of the complete network resource situation all Resource Managers synchronize with each other. Thus the Load Balancer gets the current resource situation of the complete vehicle infrastructure with aid of its local Resource Manager.

In our approach, the middleware is located on each ECU in the vehicle. Every ECU has a unique ID. The ECU with the lowest ID is the master. Thus it is responsible for the control of the entire vehicle network, and newly connected and the detaching of additional devices are discovered by its Event Manager, device information is registered by its Registry, and its Load Balancer is responsible for the evaluation of the possible migration with the aid of the local Resource Manager. If the master ECU fails a new master will be chosen with the aid of the Bully-Algorithm [18].

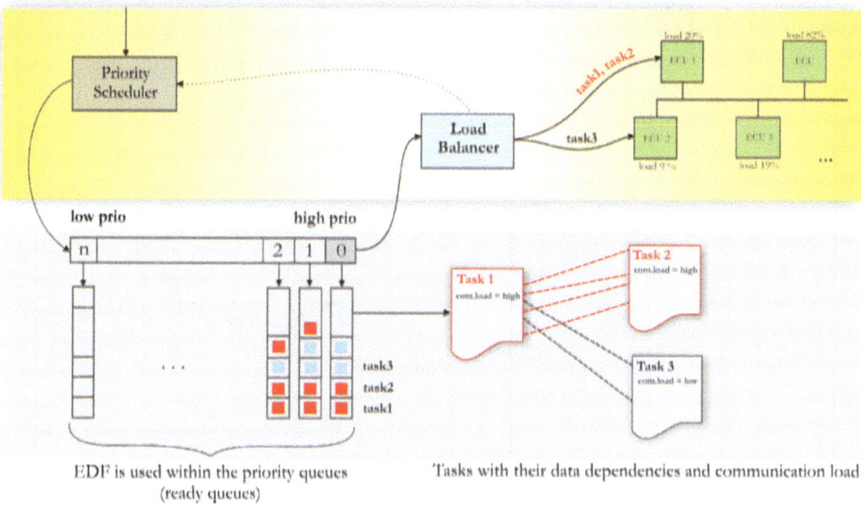

Fig. 6.2 Failure correction handling—the task migration mechanism

The failure detection if a node fails will be handled by a hardware interrupt. It initiates an error correction in our middleware. That means, to correct the error, tasks of the omitted node are migrated to other ones, which are able to execute them. In this paper we will not focus on the failure detection but on error correction. Therefore, our middleware must be able to migrate tasks. A detailed knowledge of the task characteristics is needed. It is important to know if it is a real-time task or not.

Figure 6.2 presents our approach for task migration. We assume that each task has a priority and we have a detailed knowledge about their hardware requirements. Additionally, the data dependencies between the tasks are known. As we can see from Fig. 6.2 we start with a priority scheduler. He will schedule the tasks according their priority in priority queues. That means, we have for each priority an own task queue. Within the queues the tasks are scheduled by a simple earliest deadline first (EDF) scheduler to ensure a flexible schedule [6]. Real-time (RT) tasks have a high priority. The Load Balancer works on the priority queues be-ginning from the queue with the highest up to the lowest priority. For each select-ed task a possible set of ECUs who are able to execute the task is evaluated. After that a data dependency check will be done.

That means, we look at those tasks that interact with the inspected one. In that case the interaction is weak the Load Balancer selects an ECU from the previously evaluated set of ECU and finally mi-grate the task to that one and delete the task in the priority queue. In case of a strong interaction the Load Balancer will try to avoid unnecessary busload, by selecting an ECU from the ECU set that is able to execute both tasks. Afterwards both tasks will be deleted in the priority queue. If the Load Balancer could not find a possible ECU for migration the task will be deleted from the queue with the outcome that a migration is not possible.

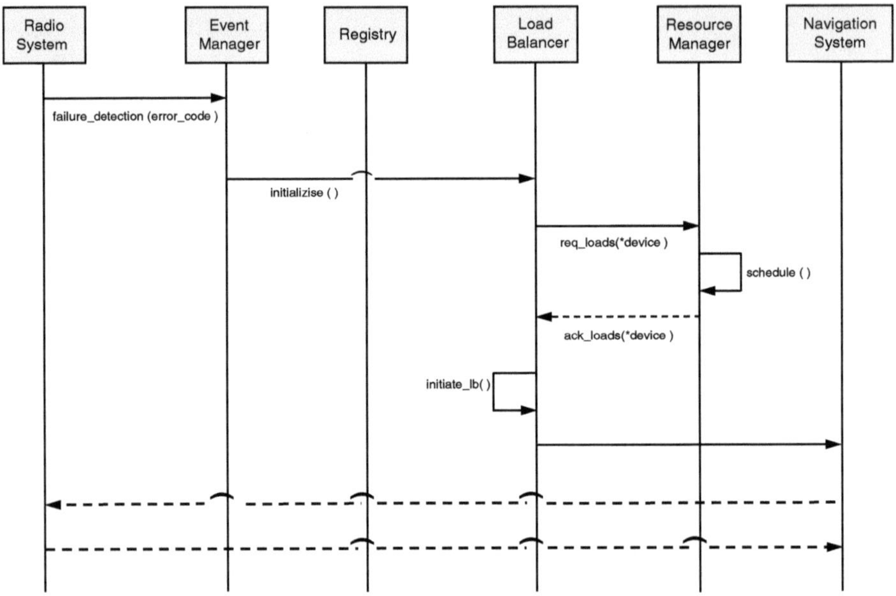

Fig. 6.3 Failure detection of the radio system

The previous paragraph give an overview of the migration, but there are still some open issues we will discuss in the following. If an ECU with more than one task running on it fails we will migrate the tasks to one or more ECUs according the classification of the tasks (see Fig. 6.2). That means tasks with high-priority will migrated first followed by the other ones. During the migration phase the tim-ing of the tasks are taken into account. After a task migration we have to decide to start the task new or from that state before the ECU fails, but how to recognize this state? Therefore, we need the context of the task. Our solution is the follow-ing, if we have a context available (e.g. store in an external flash memory of the ECU and still available) we will invoke the task with the context, otherwise not. This gives a brief overview how our middleware migrate tasks. Finally the decision which tasks are migrated is done by the Load Balancer, see Sect. 6.5.

Figure 6.3 shows a sequence diagram where a failure occurred in the radio system. We assume the tasks from the radio system can migrated to the navigation system. As we can see in Fig. 6.3 the Event Manager detects the failure of the radio system, this is done by the function failure_detection(error_code). Afterwards the Event Manager triggers the Load Balancer with the initialize() function. The Load Balancer ask for all device information from the Registry (req_loads(*device[0..n])). Then the Resource Manager runs the schedule() function to calculate all possible schedules. The Load Balancer will get the device information back from the Resource Manager with ack_loads(*device[0..n]). Finally the Load Balancer will calculate (initiate_load_balancer()) which tasks could be move from device with the failure to another one based on the information of the schedules, the load of each processing element in the car-network, the communication costs and regarding the feasibility. In our case he will decide to move tasks from the radio to the navigation system.

In the last paragraph we describe the interactions between the four tasks, which are necessary to support load balancing. Now we will discuss the internal data structure of our middleware. The Event Manager triggers the Registry and initialize the Load Balancer. The Registry itself interacts with the Resource Manager and the Load Balancer. The Resource Manager hands over the actual status of the entire system to the Load Balancer.

To perform the scheduling in the Resource Manager we can select between different scheduling strategies. They are instantiated within the scheduling mechanism class of the internal data structure.

The Registry as well as the scheduling mechanism needs information's of all tasks and devices. This is handled by the so called list class. It contains linked lists of devices and tasks and offers functions to manage the lists. As described before list offers all functions to manage the task list, but additionally functions to set the status of the tasks are needed. The status of the task is running, waiting or sleep-ing. Besides this the task manager is able to create a new task. The information of a task is stored in the data structured provided by the task control block. The parameters of the generated structure are set by the task manager with functions from the list class. The list class uses the functions from the task control block to get information's from tasks.

For the devices we have the same functions available as for the tasks. This is realized in the device control block. Each device has a list containing the task-id's that are running on the device. By setting the global variables of our middleware we can initialize the system and can set it in running mode.

6.5 Load Balancing Strategy

There are several possibilities to balance the load after an error happened inside the vehicle Infotainment network. Initiated by the Load Balancer component the new resources can be used and applications or tasks can be migrated to the additional device. In the following the cost-based load balancing strategy is briefly described. Within the cost based strategy the Load Balancer evaluates possible migration of tasks from one ECU to another. He evaluates a set of ECUs where the task could be migrated. Hence that the migration is only a useful option if

- The cost of migrating is lower than the cost of keeping tasks with their original device and
- It is feasible to migrate a task or a set of tasks from one ECU to another one (feasibility).

The cost benefit ratio for tasks of busy devices is computed which helps the Load Balancer to form the decision of whether to migrate or not. The calculation of migration costs of task is realized according to the priority list of the Most Loaded strategy. Most Loaded generates a priority list which ranks the tasks from the busiest processor. In that way the tasks with the highest priority will be mi-grated to the resources of the additional device.

Let us assume we have tasks t_i with $i = 1 \cdots n$, and the utilization of the task running on an ECU is u_i. Additionally, let U_j the maximum utilization of ECU e_j with $j = 1 \cdots m$. Then the upper bound for the utilization of an ECU e_j is:

$$\sum_{i=1}^{n} u_i \leq U_j, \tag{6.1}$$

For the communication we can make the following assumptions. Let c_k with $k = 1 \cdots r$ the communication channels in the vehicle and C_k the maximum costs a channel c_k has. Furthermore, let $m_{i,k}$ the cost task t_i produce on channel c_k. Then we can define the following bound for the communication cost a channel c_k:

$$\sum_{i=1}^{n} m_{i,k} \leq C_k, \tag{6.2}$$

Now our Load Balancer has to find an optimal balancing for all tasks within in the vehicle network regarding the utilization, communication cost and the feasibility. This can be done with integer linear programming (ILP) or other optimization methods. This is ongoing work right now and in the final version of the paper we will show some simulation results.

6.6 Algorithm and Simulation

In this section we will describe the implementation status of our middleware w.r.t. simulation and results. Our middleware was implemented in C code. We choose C, because it is more or less the language used for ECUs. Therefore, the code transfer from a PC based simulation to a real target platform doesn't need too much effort.

Within the PC based simulation we are able to parameterize our virtual software tasks and virtual ECUs with real values to achieve a software simulation of the entire system. The simulation is due to the fact that we use real values, near to the real system behavior.

The Load Balancer Algorithm is shown in Fig. 6.4. The Algorithm works as follows. For each cluster the nodes and edges are given. Furthermore, the set of tasks are stored in A. The maximum energy consumption and utilization for all nodes are also given. Now the algorithm tries to eliminate all tasks in A. Therefore, all nodes of the cluster are examined. For each task in A the dependencies to other tasks in A are checked. If two or more task have a strong dependency between each other, a new super task containing this tasks is included in a new set A'.

Algorithm LoadBalancing(cluster $\mathcal{N}$)

let $\mathcal{N} = (V, E)$ where $V = \{v_1,....v_i\}$ and $E = \{e_1,....e_j\}$

let $n \in V$ be a defect node

let $T = \{t_1,....t_m\}$ the set of tasks inside the network,
and $A \subset T$ the set of tasks of node n

let T_{v_i} be the set of tasks of node v_i, U_{v_i} the utilization of node v_i,
and P_{v_i} the energy consumption of node v_i

let U_{max} be the maximum of utilization $\forall v_i \in V$,
and P_{max} the maximum of energy consumption $\forall v_i \in V$

```
if A ≠ ∅
    for each node v_i ∈ V \ {n}
        for each task t_i ∈ A do
            choose gw(e) where gw(e) = 2
                select all tasks {t_g,t_h} from A with strong dependencies
                and set A' = A \ {t_g,t_h} ∪ {t_gh} whereas t_gh consists of t_g and t_h
                and is called super task
                    for each task t_i' ∈ A' do
                        if U_vi + u_ti' ≤ U_max ∧ P_vi + p_ti' ≤ P_max
                            if t_i' is a super task then
                                T_vi = T_vi ∪ {t_l, t_k}
                                A' = A' \ {t_i'}
                                A  = A \ {t_l, t_k}
                            else
                                T_vi = T_vi ∪ t_i'
                                A' = A' \ {t_i'}
                                A  = A \ {t_i'}
else
    LoadBalancing (neighbor cluster of N);
```

Fig. 6.4 Load balancing algorithm

After that all other tasks are included in A'. That means, A' contains super tasks and all other tasks with low dependencies. For each task of A' the utilizations and energy consumptions of each node in the cluster added by that from the task is checked. In the case we found a node for a super task, the tasks of the super tasks are added to the set of task of the node and the task is deleted in A' and the tasks of the super tasks are deleted in A. Accordingly it is not a super task, the task is added to the set of tasks and deleted in A' and A.

If A is not empty, the algorithm selects a neighboring cluster and tries to find candidates to migrate the remaining tasks.

As we figured out from the simulation that the migration time, to start a task on an ECU needs more time as our implemented scheduling and load balancing approach. Therefore, the time our middleware needs is dominated by the task migration of the underlying hardware (ECUs).

6.7 Conclusion and Outlook

We presented a middleware architecture for automotive systems that enables dynamic load balancing within the Infotainment network. The integration of load balancing is a step towards a self-reconfiguration within the vehicle and to integrate redundancy by task migration. We focus on a specific use case scenario whereby an error occurred within the vehicle network. Tasks running on the ECU with an error are migrates to another ECU by regarding the so-called feasibility, utilization and communication costs. With the help of the requirements, we de-scribed the middleware architecture and their enrichment with new services to support the distribution and exchange of tasks. Furthermore, we present briefly a cost-based load balancing strategy we will use for our approach.

Future work will be a detailed evaluation of the already existing load balancing strategies in the context of automotive systems. Additionally, the extension of existing or the development of new load balancing strategies will be done together with the implementation of the proposed architecture.

Acknowledgements This project was funded by the EU Commission within the project DySCAS (Dynamically Self-Configuring Automotive Systems). The authors would like to thank all partners of the project.

References

1. Athony R, Ekelin C, Chen C, Toerngren M, de Boer G, Jahnich I et al (2006) A future dynamically reconfigurable automotive software system. In: Proceedings of the "Elektronik im Kraftfahrzeug". Springer, Dresden, Germany
2. Athony R, Rettberg A, Jahnich I, Ekelin C et al (2007) Towards a dynamically reconfigurable automotive control system architecture. In: Rettberg A, D' mer R, Zanella M, Gerstlauer A, Rammig F (eds) Proceedings of the IESS'07. Springer, Irvine, California
3. Chaaban K, Shawky M, Crubille P (2005) A distributed framework for real-time in-vehicle applications. Intelligent Transportation Systems. Proceedings. IEEE, pp 925–929, 13–15 Sept 2005
4. Cybenko G (1989) Dynamic load balancing for distributed memory multiprocessors. J Parallel Distrib Comput 7(2):279–301. doi: 10.1016/0743-7315(89)90021-X. http://dx.doi.org/10.1016/0743-7315(89)90021-X
5. Diekmann R, Monien B, Preis R (1997) Load balancing strategies for distributed memory machines. In: Satz H, Karsch F, Monien B (eds) Multiscale phenomena and their simulation. World Scientific, Singapore, pp 255–266
6. Giorgio BC (2000) Hard real time computing systems. Kluwer, Dordecht

7. Heiss H-U, Schmitz M (1995) Decentralized dynamic load balancing: the particles approach. Inf Sci Inf Comput Sci 84(1–2):115–128
8. Hu YF, Blake RJ (1995) An optimal dynamic load balancing algorithm. In: http://citeseerx.ist.psu.edu/viewdoc/summary?doi=10.1.1.51.4963, vol DL-P-95-011
9. Hui C-C, Chanson ST (1999) Improved strategies for dynamic load balancing. IEEE Concurrency 7(3):58–67
10. Jahnich I, Podolski I, Rettberg A (2008) Integrating dynamic load balancing into the car-network. 4th IEEE International Symposium on Electronic Design, Test and Applications. DELTA pp 435–440, 23–25 Jan 2008
11. Jahnich I, Rettberg A (2007) Towards dynamic load balancing for distributed embedded automotive systems. In: Rettberg A, D'mer R, Zanella M, Gerstlauer A, Rammig F (2007) Proceedings of the IESS'07. Springer, Irvine, California
12. Jaiganesh B, Schmidt DC, Dowdy L, Othman O (2004) Evaluating the performance of middleware load balancing strategies. Enterprise Distributed Object Computing Conference. EDOC. Proceedings, Eighth IEEE International pp 135–146, 20–24 Sept 2004
13. Kim M, Choi Y, Moon Y, Kim S, Kwon O (2006) Design and implementation of status based application manager for telematics. Advanced Communication Technology. ICACT. The 8th International Conference, vol 2 pp 3–1366, 20–22 Feb 2006
14. Navet N, Song Y, Simonot-Lion F, Wilwert C (2005) Trends in automotive communication systems. Proc IEEE 93(6):1204–1223
15. Othman O, Schmidt D (2001) Optimizing distributed system performance via adaptive middleware load balancing. In: Othman O, Schmidt DC (eds) Optimizing distributed system performance via adaptive middleware load balancing. ACM SIGPLAN workshop on optimization of middleware and distributed systems (OM 2001), Snowbird, Utah, 18 June 2001
16. Othman O, Schmidt DC (2001) Issues in the design of adaptive middleware load balancing. In: LCTES '01: proceedings of the ACM SIGPLAN workshop on languages, compilers and tools for embedded systems. ACM, New York, pp 205–213
17. Ravindran B, Welch LR, Kelling C (1997) Building distributed scalable dependable real-time systems. Proceedings of the International Conference and Workshop on Engineering of Computer-Based Systems, 1997 pp 452–459, 24–28 Mar 1997
18. Stoller S (1997) Leader election in distributed systems with crash failures. In: Stoller S (eds) Leader election in distributed systems with crash failures. Technical report, Indiana University, April 1997, p 169
19. van der Zwaan S, Marques C (1999) Ant colony optimisation for job shop scheduling. In: van der Zwaan S, Marques C (eds) Ant colony optimisation for job shop scheduling. Proceedings of the third workshop on genetic algorithms and artificial life (GAAL 99). http://citeseerx.ist.psu.edu/viewdoc/summary?doi=10.1.1.53.3586

Chapter 7
Composing Adaptive Distributed Embedded and Real-Time Java Systems Based on RTSJ

M. Teresa Higuera-Toledano

Abstract Distributed safety-critical systems that must provide continuous service (e.g., air-traffic control systems or enterprise and financial applications) require dynamic updates. Dynamic software modification is a useful capability because the application is able to update it to fix bugs and add new features without requiring a stop and a restart. During the last years (i.e., since 1998) the Java technology has been extended in order to provide the timelines required by embedded real-time distributed systems (i.e., RTSJ, DRTSJ, SCJS). However neither of these Java extensions considers dynamic characteristics allowing the system to be adapted to the changing environment. In this Chapter, we overview Java solutions considering real-time requirements in service-oriented component models allowing dynamic updates (e.g., OSGi).

7.1 Introduction

By *real-time systems*, we understand computer programs that must execute with time constraints. Hence, *timeliness* parameters are part of the system's behavioral requirements. A real-time system can be characterized by its execution model, its scheduling algorithm, and its synchronization protocol. Real-time and embedded applications cover an extremely wide variety of domains, each with stringent requirements. The interdependence between functional and real-time semantics of real-time software makes the maintenance especially difficult. In addition,

M.T. Higuera-Toledano (✉)
Universidad Complutense de Madrid, Ciudad Universitaria, Madrid 28040, Spain
e-mail: mthiguer@dacya.ucm.es

M.T. Higuera-Toledano et al. (eds.), *Self-Organization in Embedded Real-Time Systems*, DOI 10.1007/978-1-4614-1969-3_7,

embedded software systems are not portable as they depend on the particular underlying operating system and hardware architecture.[1]

The breadth of domains covering of real-time and embedded application results in a deep diversity of requirements, including highly precise timing characteristics, small memory footprints, flexible sensor and actuator interfaces, and robust safety characteristics. Since many embedded and real-time applications can be categorized as mission-critical or safety-critical, where even human life is sometimes at stake, a major part of the cost of creating such real-time and embedded applications is spent in the complex software development, integration, verification, and validation of the resulting systems.

Therefore it is essential for the production of real-time embedded systems to take into account modern languages, tools, and methods allowing exploit beneficial effects such as resource management and highly robust structures, which simultaneously provides high software productivity and easier maintenance. Moreover, processing in embedded and real-time application domains often involves distributed processing (e.g., automobile braking or aircraft flight control) in which communication and synchronization and their resulting integration present additional challenges. The Java programming language has become an attractive choice because of its safety (e.g., it has not pointers), productivity and relatively low maintenance costs.

The last years have been a rapid progress in processor and sensor technologies in both single-node and distributed applications. This progress, combined with the expanding diversity of application domains, is placing enormous demands on the facilities that the Java run-time environment must provide. The Java technology has been extended in order to provide the timelines required by real-time systems:

(i) In 1.998 the *Java Community Process* approved the first JSR (i.e., the JSR-001) dedicated to the *Real-Time Specification for Java* (RTSJ) [13], which is an extension to Java adding real-time programming constructs and constraints to the Java environment.

(ii) On Mars 2000, the JSR-050 [14]) started as a RTSJ-based profile dedicated to the *Distributed Real-Time Specification for Java* (DRTSJ). This specification adds time predictable *end-to-end* characteristics to RTSJ in order to support distributed real-time Java systems. Since this specification has been inactive for a long time, in February 2011 a new *Expert Group* has been created. The goal of this new group is to extend RTJS with a real-time implementation of the *Java's Remote Method Invocation* (RMI).

(iii) On August 2005, just after approved JRS-001 a new RTSJ version started (i.e., the JSR-282) [17]. This new RTSJ proposal addresses some lacks and enhancements that have been discovered and requested in the first RTSJ version (i.e., the JSR-001).

[1]This holds even stronger for (hard) real-time systems since the guarantee of real-time behavior can only be given for a concrete hardware platform.

(iv) The real-time community has initiated, on July 2006, another JCP standardization process (i.e., the JSR-302 [18]) dedicated to *Safety-Critical Java Specification* (SCJS), which defines those capabilities needed to use Java technology to create safety critical applications. SCJS is an RTSJ-based profile, which supports the development of programs that must be certified. This specification includes an annotation set to constrain the behavior of programs and the specific rules to check statically the semantic of the programs. Given that the application validation is top priority in safety-critical systems, it will require a transformation from Java *bytecodes* to target machine representation prior to certification. Additionally, SCJS will specifically identify all classes and methods on which a safety critical application can depend at runtime (i.e., without classes and methods not used by the application); so that the DO-178B dead code elimination requirements can be supported.

The increasing capabilities of modern microprocessors and micro-controllers greatly increase their applicability to complex environments requiring dynamic reconfiguration or adaptation. As a consequence, the interest in technology allowing the software to be adapted dynamically to its environment has increased considerably in the past few years. Adaptive computing systems are now-days driving in two directions[23]:

(i) The high development of *ubiquitous* computing (i.e., *ubicomp*), which deals with the model of human-computer interaction, pervasive computing, or ambient intelligence (e.g., mobile computers and wireless networks).
(ii) The emergence of *autonomic* computing, which deals on self-X characteristics (e.g., financial networks and power grids). This kind of systems must be able to manage and protect themselves (e.g., these systems are tolerant to hardware failures).

Software adaptation can be achieved by using two implementation approaches: *(i)* by modifying the behavior program variables (i.e., *parameter adaptation*), or *(ii)* by modifying the algorithm or the system structure (i.e., *compositional adaptation*). While parameter adaptation can only tune the strategy of an application by tuning its parameters, compositional adaptation allows an application to incorporate new algorithms during all its life cycle (i.e., it enables reconfiguring the application during its execution phase).

The dynamic software reconfiguration (e.g., self-modifying code or overlays-based execution) dates from the first computing generation. Given the increasing interest of research in this topic, now days several tools and technologies allow us to address this problem. However, dynamic reconfiguration is still challenging for embedded real-time control software systems because:

- Component-based software engineering[2] which is the main approach in order to deal with the increasing software complexity issues, among them several benefits are *reuse*, *abstraction*, *division of labor* and *ease of maintenance*.

[2]Software engineering dealing with software components, reusable software elements with a specification.

- It is not possible to anticipate the effects of dynamic modifications which are not planned at design time.
- It can affect predictability, which is inadmissible in safety-critical systems.
- It can also add unsafe modules, which can bring the application to a wrong state, or even make the whole platform crash. This flexibility has a cost (i.e., safety can be compromised).

This chapter presents some significant research work on the use of components in Java embedded real-time systems, based on their innovative ideas and novel contributions on this topic. We present three groups of related works:

- First, we address works dealing with the integration of components in Java-based real-time systems.
- The second group of work deals with components in real-time and distributed Java systems.
- The third group addresses the work related to safety critical Java components.

7.2 Components-Based Solutions

Compositional techniques are the result of years of research in software engineering. Structuring systems as interacting components facilitates their design, analysis and construction process, and deals with scalability, evolution and complexity. Regarding component-based real-time systems, the most important principle to be considered is that non-functional properties such as timeliness and testability must not be affected by the system integration [21]. The last years have been prolific in research works focusing on how to introduce real-time requirements in the component models, some of them are also RTSJ-compliant.

Some real-time applications (e.g., home automation, pervasive environments and even business contexts) require some forms of dynamic support. The dynamic reconfiguration enables an OSGi [1] application to be modified at run-time, without turning of the application. Taken into account the high availability of real-time systems, this feature is particularly interesting because it allows an application to evolve, adapt and be maintained without affecting the application availability. Also, it is implemented in Java. However, it does not have support for real-time applications. Dynamic adaptation affects the *Worst Case Execution Time* (WCET) analysis because runtime updates in a component can modify the WCET of all the threads using the updated component. Resource reservation may cause overload situations given the impossibility to predict the number of components in the system. Moreover, it may install more components than it is possible to guarantee resources for; consequently, real-time threads can miss their deadlines affecting the time requirements of components.

Taken into account the challenge to integrate the hard, soft, and non real-time components that these complex systems require, RTSJ represents an attractive choice. However, RTSJ at the implementation level is error-prone because it presents non-intuitive rules and restrictions (e.g., the *single parent rule* or the assignment rules preventing dangling memory references [5, 11]). On the other hand, one of the most used techniques for ensuring predictable behavior of real-time applications is the anticipation of resource reservation. However, a solution to resource management and negotiation has not yet been addressed by RTSJ. How resource management is treated by existing component-based solutions are described in the following.

7.2.1 Components for RTSJ

This proposal focuses on the benefit of separating functional and non-functional requirements in the design and implementation stages of RTSJ applications [27]. This solution makes a distinction between *Active* components (i.e., containing a control thread) and *Passive* components (i.e., representing services). For active components, the framework manages their Lifecycle and generates code that activates their functionality.

The RTSJ introduces memory regions and allows the implementation of real-time compliant garbage collectors to be run within regions except within those associated with hard timing constraints. RTSJ makes a distinction between three main priority levels: *(i)* critical tasks that cannot tolerate preemption latencies because the GC, *(ii)* high-priority tasks that cannot tolerate GC unbounded preemption latencies, and *(iii)* low-priority tasks that tolerate the collector pauses. In RTSJ, critical tasks are instances of the `NoHeapRealtimeThread` class, high-priority tasks are instances of the `RealtimeThread` class, and low-priority are instances of the `Thread` class.

In the above context, a non-real-time collector introduces unpredictable pauses that are not tolerated by critical and high-priority tasks. Real-time collectors eliminate this problem but introduce a high overhead. Also, bounded pauses are not tolerated by critical tasks. An alternative approach is to use regions within which allocation and deallocation is customized. Each memory region is then managed so as to embed objects that are related regarding associated lifetime and real-time requirements. Such a facility is enabled by the RTSJ specification (see Fig. 7.1):

The `MemoryArea` abstract class supports the region paradigm [10] in the RTSJ specification through three kinds of regions (see Fig. 7.1): (*i*) immortal memory, supported by the `ImmortalMemory` and the `ImmortalPhysica-lMemory` classes, that contain objects whose life ends only when the JVM terminates; (*ii*) (nested) scoped memory, supported by the `ScopedMemory` abstract class that enables grouping objects having well-defined lifetimes and that may either offer temporal guarantees (i.e., supported by the `LTMemory` and `LTPhysicalMemory` classes) or not (i.e., supported by the `VTMemory` and `VTPhysicalMemory` classes) on the time taken to create objects; and (*iii*) the conventional heap, supported by the `HeapMemory` class.

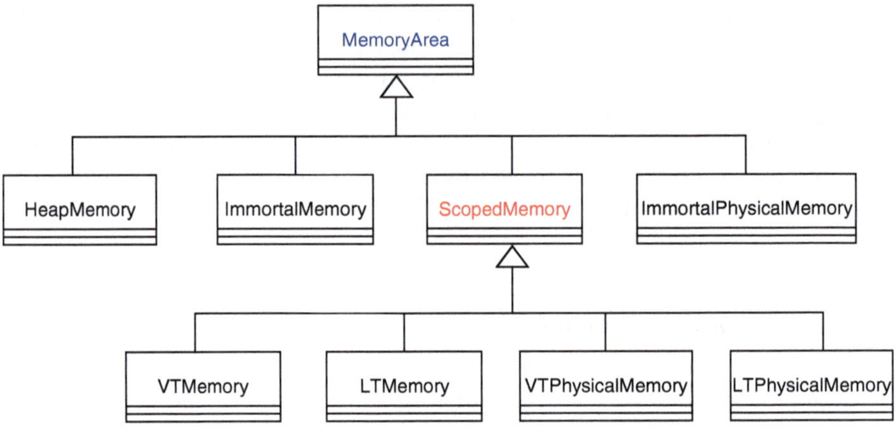

Fig. 7.1 The `MemoryArea hierarchy in RTSJ`

Regarding the *Components for RTSJ* solution, the *ThreadDomain* component represents the `Thread` hierarchy (i.e., `RealTimeThread`, and `NoHeap-RealTimeThread`) encapsulating all the active components with the same properties (e.g., thread-type or priority). The *MemoryArea* component represents the `MemoryArea` hierarchy of RTSJ (i.e., `ImmortalMemory`, `ScopedMemory`, and `HeapMemory`) subclasses, encapsulating all subcomponents that are allocated in the same memory area.

RTSJ imposes restricted assignment rules that keep possibly longer-lived objects from referencing objects that may have a shorter life. These restrictive memory-management use of scoped memory regions [5, 8] impose also several restrictions on the composition process (e.g., a *NHRT ThreadDomain* should not encapsulate a `HeapMemory` object). After deploying all components, the adherence to RTSJ must be verified. The functional architecture is obtained as a combination of the *ThreadDomain* and the *MemoryArea* components, which allows us to change real-time requirements without modifying the functional architecture. The verification process ensures that compositions violating RTSJ rules will be refused.

7.2.2 RTComposer

The *RTComposer* framework [2] is also an RTSJ-based solution, which uses formal specification of scheduling constraints. The scheduling specifications can be given as periodic tasks, using temporal logic, or as an omega-automata. Components can be added dynamically, and non-real-time components are allowed. Each component can be specified and analyzed in a platform-independent way, and its performance can vary with changing load or processing speed.

Each component is scheduled in a flexible way, which may vary according to dynamic conditions (e.g., varying load, platform capabilities, or component configurations), and has a clearly specified interface including its used methods (e.g., for sensing, computation, and actuation) and its requirements are given as a set of *macro-schedules*. Each *macro-scheduler* composes the specifications of all components along with the platform specification constrains, which methods can be executed within a *micro-scheduler* implemented on top of a native priority-based scheduler.

7.2.3 *RT-OSGi*

The *RT-OSGi* solution [29] allows develop service-oriented applications in an RTSJ environment. This solution exploits the power of both, dynamic configuration from the OSGi framework and RTSJ priority-based scheduling. *RT-OSGi* provides an admission control protocol to the OSGi framework solving the unbounded dynamism problem. This solution also provides a priority assignment approach to support temporal isolation. The combination of both, the admission control protocol and temporal isolation, allows install components into the system guaranteeing their resources and ensuring safety update. However, *RT-OSGi* does not deal with memory consumption.

Temporal isolation is achieved by implementing execution-time servers [31] as CPU time reservation mechanism, which requires hierarchical scheduling [7]. The general idea is to use the RTSJ `ProcessigGroupParameters` (PGP) class and application level cost enforcement. For each component a PGP object is created, and it is assigned a CPU time bugged per period. A real-time thread can use no more than the assigned PGP CPU time budged per period. The JVM monitors the CPU time used by real-time threads. Whether the budged assigned to a PGP is used, the real-time threads in the group lower their priorities to a background priority, and raises them to the original priorities whether the group budged is replenished at starting the PGP period.

7.2.4 *Freezing Architecture*

The *Freezing Architecture* approach [9] considers an application as a set of states, each of which corresponds to a given architecture of the application. Transitions between states correspond to changes in an application component at run-time (i.e., arrival, departure or update). As a consequence, real-time constraints are guaranteed by defining state transition rules between states in order to respect the application timing constraints. We must define rules for the transitions between states.[3] In order to ensure real-time requirements, once a system enters a real-time state, no modifications are performed until it returns to the corresponding non-real-time stats. The system holds all the architecture changes until it leaves the critical state.

[3]In the OSGi platform, these transitions are performed automatically.

This approach has been implemented by using an iPOJO[4] *Dependency Handler*, which provides data on service dependencies and may intercept modifications in the application Lifecycle [9]. All the handlers in the system are connected to a *Real-time State Manager* (RTSM), which verifies the platform state before a modification (i.e., before adding, modifying or removing a service dependency) on its architecture. The thread responsible for architectural modification is blocked until the system reaches a non real-time state. This approach allows a global context of the platform avoiding components interfere with the architecture. In addition, real-time components are directly connected to the RTSM in order to switch the system state between real-time and non real-time.

7.2.5 Discussion

Compositional adaptation can be achieved by using three different approaches: *(i)* concern separation (e.g., aspect oriented programming), *(ii)* computational reflection (e.g., meta object protocols), and *(iii)* component-based design (e.g., software components as Java Beans). These three technologies are based in an indirection level.

Component-oriented frameworks, as the *Components for RTSJ* approach [27], can abstract these complexities by adding a middleware layer. This approach develops a component model allowing fully separation of functional and non-functional requirements. Regarding RTSJ, this solution may lead to potential memory assignment issues, due to its rules which prevent dangling memory references. Furthermore, the fact that RTSJ class objects are allocated within the `ImmortalMemoryArea` and the explicit object allocation in this memory region complicates class unloading generating memory leaks.

Resources allocated to a component must be isolated from the resources of the rest of the components in the system. In particular, a hard problem is around computational resources (e.g., CPU time and memory use). To guarantee that each real-time thread meet its deadline, each component requires its resource reservation (i.e., CPU time and memory consumption). Resource management can be studied from three points of view:

- The ability to track resource usage (*resource accounting*).
- The ability to reclaim the resources of a component when it terminates (*resource reclamation*).
- The ability to allocate resources to a component (*resource allocation*).

[4]An iPOJO is a runtime service-oriented component model which can be used to develop applications over the OSGi service platform. Besides managing dynamic service interactions *iPOJO* also provides an extensibility mechanism to manage other non-functional properties as persistency, security, autonomic management, and events.

Each component must have an expected limits for both CPU utilization and memory consumption. These limits can be previously obtained via profiling and measurement on a determined targeted platform. At run-time the resource needs of each component must be guaranteed, which must enforce the established resource budgets.

RT-OSGi [29] achieves time-based problems by establishing priority range mapping rules, which allows global resource budget enforcement for components, and ensures component real-time contracts. However this solution deals only with CPU time; memory consumption and RTSJ memory areas has not been still addressed. *RT-OSGi* establishes the bounds placed on the number of components in the system, ensuring that new installed components meet their deadlines without disturbing the other components to miss their deadline. This solution provides temporal isolation by monitoring PGP CPU time consumption.

As the *RT-OSGi* approach, the *Freezing Architecture* solution [9] uses resource usage as one of the criteria for the admission protocol, but as different about it, this approach uses *Service Level Agreement* (SLA). As a component-based real-time systems are designed to control the whole life-cycle, this type of systems can adjust their-self to operate without affecting real-time contracts even whether the run-time changes. One limitation of the *Freezing Architecture* approach is dealing with physical devices or network services.

In order to guarantee the system real-time requirement in presence of run-time changes, we require some global information and general control interfaces. This global information allows also a global view of the real-time context, which becomes imperative. Scheduling analysis allows assign priorities within components, but not across components, which can cause problems like deadlocks, starvation, missed deadlines, and incorrect results due to precedence violations.

7.3 Distributed Real-Time Java Issues

Much research in adaptive systems focuses on *middleware* (i.e., the software levels between the application and the operating systems or the network protocols). Most adaptive middleware uses the object-oriented paradigm (i.e., DCOM, CORBA, and Java RMI) and work by using messages passing supported by the *client-server*) architecture model.

Network bandwidth is a resource not traditionally treated in real-time systems. However, network bandwidth is a highly important resource in real-time Java systems, due to the network and distributed nature of the Java environment. A well-known example is RT-CORBA [30], where the middleware is divided into four layers:

1. The *host-infrastructure middleware* level, under the operating system, provides an API that hides the heterogeneity of the platform (i.e., hardware devices, operating system and network protocols).

2. The *distribution middleware* level provides distributed programing abstractions (i.e., *remote objects*) enabling the development of distributed applications.
3. The *common middleware* level adds services such as fault tolerant, secure, or persistence.
4. The *domain-specific middleware* allows to support a particular class of applications (i.e., real-time).

The main function of middleware is to hide the both resource distribution and platform heterogeneity, given an abstraction level to the application programmer. Thus this abstraction level is the ideal place to add adaptive features.

DRTSJ (i.e., JSR-50 [14]) will allow Java programmers to write programs that meet their real-time constraints in distributed systems, and defines three integration levels, each one requiring some support from the underlying system and offering some benefit to the programmer:

1. *L0* is the minimal integration level, it considers the use of RMI without changes;
2. *L1* is the first real-time level, it considers real-time properties in the RMI; and
3. *L2* offers a transactional model for RMI, and introduces the concept of *distributable* real-time threads, which transfer information across the nodes.

7.3.1 Profiles for RT-RMI

The *RT-RMI* solution [6, 32, 34] is a profile definition for time-predictable RMI in RTSJ, which also addressed issues related to improving the efficiency. Since different applications may require different profiles, three different environments are considered:

- **Safety critical systems** present a correct and highly deterministic behavior (i.e., for hard real-time systems); must support systems where deadline misses can cost human lives or cause fatal errors. The *RMI-HRT* profile [34] defines the requirements of this level (see Sect. 7.4).
- **Business-critical systems** present efficiency and robustness as extra-functional requirements (i.e., for soft real-time systems). This profile must support systems whose anomalous behavior may have financial costs. The corresponding profile, named *RMI-QoS*, has also *end-to-end* constraints but does not requires *Worst Case Response Time* (WCRT) analysis, allowing certain deadline misses whether they fit in the *Quality-of-Service* (QoS) patterns [6].
- **Flexible business-critical systems** deal with flexibility (e.g., multimedia systems and future ambient intelligent systems). Some features of RTSJ (e.g., *Asynchronous Interrupt Exception* (AIE), or *Asynchronous Event Handling* (AEH)) are not considered in the framework, while considering a real-time GC. The corresponding profile considers a support based on OSGi, with bundles that manage certain resources [25, 32].

There are four features defining each profile:

- *(i)* The computational model, which relates to the model describing the *end-to-end* predictable behavior,
- *(ii)* The adaptation required on RMI, which defines specific classes and hierarchies for specific domains supported inside by the middleware core,
- *(iii)* The concurrency model in the remote invocation, which refers to the rules governing the remote invocation at both client and server side, and
- *(iv)* The general memory considerations, which relates to the memory requirements used during the application execution.

7.3.2 *Compadres*

The *Compadres* [12] solution presents a component framework for distributed real-time and embedded systems, offering several advantages, among them:

- *(i)* It abstracts the memory management complexity inherent to RTSJ, since the compiler that automatically generates the scoped memory architecture for components,
- *(ii)* It provides communication among components through strongly-typed objects,
- *(iii)* It handles the communication between the components, which are assembled by means of connecting ports, and
- *(iv)* It deals with performance and efficiency.

This approach is based on a hierarchical component model where each component is allocated either within a `SopedMemoryArea` object or within and `ImmortalMemoryArea`. Therefore, it is not possible to use regular threads[5] (i.e., non-real-time threads). As a consequence, this component-model loses the compliance with the standard Java (i.e., non-real-time Java) having gone down an important feature of the RTSJ. Moreover, the communication model has been reduced to event-oriented interactions between components. *Compadres* also proposes a design process of real-time applications. However, the complexity of designing real-time systems is not really addressed.

7.3.3 *Event Manager Component*

The *Event Manager Component* approach [22] extends and generalizes AEH to provide component managing of distributed events. This solution presents configurable

[5]Regular threads allocates objects within the application heap.

CORBA-IDL interfaces allowing AEH. The model has a global view of the system from the event happening with the release on a processor, treating the event handlers as servers and the processors as resource allocations.

The allocation or mapping process consists to assign individual event handlers to servers and servers to processors. The component uses the description of the events relations to generate a state automaton used to release their handlers. The feasibility analysis processes at a component level and takes into account the events relations and the association of the servers with the processors (i.e., *affinity*). As in [20], this approach uses automata-based models in order to explore and verify their properties.

7.3.4 RTSJ-Container

The component-based approach presented in [35] integrates the RTSJ services based on a resource reservation model under the *Java Platform Enterprise Edition* (J2EE), which has been developed under two JCPs; JSR-058 (i.e., version 1.3) [15] and JSR-151 (i.e., version 1.4) [16]. J2EE uses middle-tier server solutions to reduce the complexity of clients and application services. The *Enterprise Java Beans* (EJB) container is a basic J2EE-based profile supporting components. EJB supports services such as transaction, persistence, and security. However EJB does not include real-time facilities. The goal of the *RTSJ-Container* approach is to avoid the problems of incorrect use of RTSJ APIs and simplifies the component.

Each component has associated: *facets* (i.e., interfaces synchronously or asynchronously used by clients), and *receptacles* (i.e., interfaces synchronously and asynchronously used by the component). Each component negotiates the arrival-pattern and latency for its *facets* and *receptacles*. Here, the process of negotiation establishes the temporal properties of component execution. The container implements the negotiation associated to the *facets* as resources available and reserves the resources. The negotiations are expressed in terms of the quality attributes that define the arrival-patterns and latencies of method invocations and memory usage that support the component.

The `RTObject` interface, is a common interface for all real-time components, which includes methods for the negotiation and other system services as: *(i)* the arrival-pattern invocation that the component can support (e.g., periodic, aperiodic, group-server), *(ii)* the specific parameters of each pattern (e.g., minimum period, maximum number of concurrent executions), and *(iii)* the allowed latency (e.g., deadline) for the responses.

7.3.5 Discussion

In general, the behavior and resource consumption of a component-based distributed real-time embedded can be determined such as:

- *(i)* The structure of the component assembly.
- *(ii)* The deployment of the components of hardware nodes.
- *(iii)* The properties of each component related to behavior and resource consumption.

Some mechanisms related to the RMI present a source of indeterminacy and must be removed from a real-time platform (e.g., class downloading), while others can be integrated, but their behavior should be clarified (e.g., registry, AIE, and AEH). Others are complex and introduce high overhead (e.g., object serialization [33] and distributed GC).

In general, proposed solutions assumes a TCP/IP network with a RSVP management infrastructure as it is described in [24] (e.g., *Profiles for RT-RMI*) . Since the compliance with the standard Java is considered one of the biggest advantages of the RTSJ, it should keep. However, the *Compadres* component model misses this feature. The *RTSJ-Container* introduces a solution for the integration of QoS basic services, such as resource reservation and negotiation. The EJB containers implement some basic negotiation algorithms and isolate the business components from reservation services. However, they have limitations in their application in real-time and reliable systems, because they do not include services of predictability and dependability.

7.4 Components for High-Integrity Systems

The SCJS profile (i.e., JSR-302 [18]) reduces, restricts and simplifies the RTSJ programming model to be suitable of verification and validation of safety-critical codes (e.g., for aerospace vehicles or air-traffic control systems, or enterprise and financial applications). Safety-critical systems are distinguished by stringent requirements (e.g., space, time predictability, dependability, and safety) at both process and product level. Such systems must pass a verification and validation process. The rising complexity of functional requirements and the concurrent demand for higher quality greatly increases the development cost and effort required by these systems.

The new standards emerging for SCJS present the following characteristics:

- *i)* No garbage collection.
- *ii)* Safe stack-allocation.
- *iii)* Very small memory footprint.
- *iv)* Simple run-time environment.
- *v)* Efficient throughput.

In order to provide static correctness guarantees to the RTSJ memory model, a number of research works restrict real-time *profiles* in which the complexity of scoped regions is highly reduced, and their flexibility limited. Considering the cost of failure in safety-critical systems, the effort of adopting static memory management in such as systems can be justified.

The complexity of software in safety-critical systems have reached the point where new techniques are needed to ensure system dependability while improving productivity. One possible approach is to use a component-based platform, which

- *(i)* Enables the system to be created from separate components,
- *(ii)* Restricts component interactions preventing fault propagation, and
- *(iii)* Allows compositional well-known properties.

In order to have an OSGi based safety-critical system, each component must specify it worst case utilization for both CPU and memory, which can be achieved using static analysis or programming annotations, or both. In order to that, instead of global static analysis to automatically determine worst case resource needed, a local static analysis support is required.

In the last few years, component-based engineering encourages systematic software reuse. Regarding the integration of this new design paradigm and SCJS technologies, the ideal solution consists to provide an OSGi-profile supporting real-time capabilities for high-integrity environments. Following, we present the existing component-based solutions for safety critical systems.

7.4.1 RMI-HRT Profile

In the context of the DRTSJ (i.e., JSR-050), the *Profiles for RT-RMI* solution [34] presents a profile for safety critical systems (see Sect. 7.3), which is based on preemptive priority-driven scheduling and requires the use of priority ceiling protocols (i.e., the *Safety Critical Systems Profile*). This profile considers two execution phases: (i.e., *initialization* and *mission*), defines the use of the `ImmortalMemoryArea` for initialization and `ScopedMemoryArea` in a mission, and introduces additional constraints in the RMI model [28].

7.4.2 Components for SCJS

A component-based development approach for high-integrity real-time systems has been studied in [26], which is constructed around four fundamental ingredients:

- *(i)* A component model,
- *(ii)* A computational model,
- *(iii)* A programming model, and
- *(iii)* An execution platform.

The computational model is particularly studied, because comprises architectural concerns, and is the bridge between the component model and the static analysis theories used in safety-critical systems.

7.4.3 PERC and OSGi

The *PERC Raven* solution from Aonix[6] supports hard real-time and certifiable safety-critical virtual Java machine, providing a working environment allowing to develop, validate and verify complex applications in an extensive range of engineering sectors such as aerospace, defense, automotive, transportation, telecommunications, electronics, and medical environment. There has been an attempt to integrate PERC[7] [3, 4] and an event-based profiling of OSGi.

7.4.4 Discussion

The platform must guarantee the worst case resource budgets for all the components in the system; missed deadlines cannot be tolerated. To guarantee CPU and memory budgets, classes must be allocated in a stack-based way, and each class may be accessed only by calling methods and descendent threads. In order to add real-time capabilities to a new OSGi-profile supporting safety critical Java systems, we can take as reference the *Ravenscar-Java* profile [28], which selects a subset of RTSJ features allowing us to implement high-integrity real-time programs. At this point there are two possible solutions based on: *(i)* refactoring the OSGi specification to create a high-integrity version, and *(ii)* creating a real-time version of the OSGi framework by using code instrumentation techniques.

Some initial steps towards building a platform that combines component-based software construction with hard real-time operating system services are given in [19]. However, this platform is not Java-based and specifically deals with the *CORBA Component Model*[8] (CCM) and the ARINC-653 platform services.

7.5 Summary and Conclusions

Due to its interaction with the real-world, some real-time embedded systems must be dynamically adaptive, that is, they must be capable of being modified and updated at runtime. Software runtime adaptations may be parametric or compositional functions; the former means modifications in the system variables, while the latter specifies addition or removal of system components. In this paper, we have given an

[6] http://www.aonix.com/pdf/PERC-Raven_e.pdf.

[7] *PERC Ultra* is typically used for large scale applications that need to use the rich features of *Standard Edition Java* libraries, coupled with predictable performance. *PERC Pico* is specifically designed for small footprint applications with fast execution, and low level device access, and *PERC Ultra SMP*, provides solution for multi-core, multi-processor systems.

[8] http://ditec.um.es/~dsevilla/ccm/.

overview of Java solutions considering real-time requirements (i.e., RTSJ, DRTSJ, and SCJS) in service-oriented component models.

Since even critical real-time software must be updated due to the environment changes or maintenance and cannot have its execution interrupted, dynamic adaptive techniques with flexibility capabilities taken into account possible changes in its environment without powering off the whole system are the ideal solution. However, runtime adaptive software and real-time software both are disjoint sets due to the conflict between predictability and flexibility.

The combination of both Java approaches (i.e., components and RTSJ-based technologies) is a good solution to reduce the complexity of real-time Java programs.

Acknowledgements This research was supported by Consejería de Educación de la Comunidad de Madrid, Fondo Europeo de Desarrollo Regional (FEDER) and Fondo Social Europeo (FSE), through Research Program S2009/TIC-1468, and by Ministerio de Educación y Ciencia, through the research grant TIN2009-.07146.

References

1. Alliance O (2012) Osgi service platform release 4. URL http://www.osgi.org/Specifications/HomePage. Accessed on 2012
2. Alur R, Weiss G (2008) Rtcomposer: a framework for real-time components with scheduling interfaces. In: International Conference on Embedded Software (EMSOFT), pp 159–168. Accessed on 2012
3. Aonix (2010) Aonix perc. URL http://www.aonix.com/perc.html. Accessed on 2012
4. Atego (2010) Atego aonix perc. URL http://www.atego.com/products/aonix-perc. Accessed on 2012
5. Bollella G, Reinholtz K (2002) Scoped memory. In: IEEE International Symposium on Object/Component/Service-Oriented Real-Time Distributed Computing (ISORC), pp 23–25
6. Briones JF, de Miguel MA, Alonso A, Silva JP (2009) Quality of service composition and adaptability of software architectures. In: IEEE International Symposium on Object/Component/Service-Oriented Real-Time Distributed Computing (ISORC), pp 169–173
7. Davis RI, Burns A (2005) Hierarchical fixed priority pre-emptive scheduling. In: IEEE Real-Time Systems Symposium (RTSS), pp 389–398
8. Dvorak D, Bollella G, Canham T, Carson V, Champlin V, Giovannoni B, Indictor M, Meyer K, Murray A, Reinholtz K (2004) Project golden gate: Towards real-time java in space missions. In: IEEE International Symposium on Object/Component/Service-Oriented Real-Time Distributed Computing (ISORC), pp 15–22
9. Escoffier C, Hall RS, Lalanda P (2007) ipojo: an extensible service-oriented component framework. In: IEEE SCC. IEEE Computer Society, Silver Spring, pp 474–481
10. Gay D, Aiken A (1998) Memory management with explicit regions. In: Programming Language Design and Implementation (PLDI), pp 313–323
11. Higuera-Toledano MT (2006) Analyzing the memory management semantic and requirements of the real-time specification of java jsr-0000001. In: IEEE International Symposium on Object/Component/Service-Oriented Real-Time Distributed Computing (ISORC), pp 419–423
12. Hu J, Gorappa S, Colmenares JA, Klefstad R (2007) Compadres: A lightweight component middleware framework for composing distributed real-time embedded systems with real-time java. In: Middleware 2007, ACM/IFIP/USENIX 8th International Middleware Conference, pp 41–59

13. JSR-000001, S.J.C.P. (1998) Real-time specification for java. URL http://www.jcp.org/en/jsr/detail?id=1. Accessed on 2012
14. JSR-000050, S.J.C.P. (2000) Distributed real-time specification. URL http://www.jcp.org/en/jsr/detail?id=50. Accessed on 2012
15. JSR-000058, S.J.C.P. (2000) JavaTM 2 platform, enterprise edition 1.3 specification. URL http://jcp.org/en/jsr/detail?id=58. Accessed on 2012
16. JSR-000058, S.J.C.P. (2001) JavaTM 2 platform, enterprise edition 1.4 (j2ee 1.4) specification. URL http://jcp.org/en/jsr/detail?id=151. Accessed on 2012
17. JSR-000282, S.J.C.P. (2005) Rtsj version 1.1. URL http://www.jcp.org/en/jsr/detail?id=282. Accessed on 2012
18. JSR-000302, S.J.C.P. (2006) Safety critical javatm technology. URL http://www.jcp.org/en/jsr/detail?id=302. Accessed on 2012
19. Karsai ADG, Kereskenyi R, Mahadevan N (2010) A real-time component framework: Experience with ccm and arinc 653. In: IEEE International Symposium on Object/Component/Service-Oriented Real-Time Distributed Computing (ISORC)
20. Kim M, Wellings AJ (2010) Efficient asynchronous event handling in the real-time specification for java. ACM Trans Embedded Comput Syst 10(1)
21. Kopetz H (1997) Component-based design of large distributed real-time systems. Contrl Eng Pract 6:53–60
22. Masson D, Midonnet S (2010) The design of a real-time event manager component. In: International conference on New Technologies of Distributed Systems (NOTERE), pp 291–296
23. McKinley PK, Sadjadi SM, Kasten EP, Cheng BHC (2004) Composing adaptive software. IEEE Comput 37(7):56–64
24. de Miguel MA (2001) Solutions to make java-rmi time predictable. In: IEEE International Symposium on Object/Component/Service-Oriented Real-Time Distributed Computing (ISORC), pp 379–386
25. de Miguel MA, Higuera-Toledano MT (2004) Runtime management of quality specification for qos-aware components. In: EUROMICRO, pp 84–91
26. Panunzio M, Vardanega T (2009) On component-based development and high-integrity real-time systems. In: IEEE International Conference on Embedded and Real-Time Computing Systems and Applications (RTCSA), pp 79–84
27. Plsek A, Loiret F, Merle P, Seinturier L (2008) A component framework for java-based real-time embedded systems. In: Middleware, pp 124–143
28. Puschner PP, Wellings AJ (2001) A profile for high-integrity real-time java programs. In: IEEE International Symposium on Object/Component/Service-Oriented Real-Time Distributed Computing (ISORC), pp 15–22
29. Richardson T, Wellings A (2010) An admission control protocol for real-time osgi. In: IEEE International Symposium on Object/Component/Service-Oriented Real-Time Distributed Computing (ISORC)
30. Schmidt DC (2002) Middleware for real-time and embedded systems. Comm ACM 45(6): 43–48
31. Strosnider JK, Lehoczky JP, Sha L (1995) The deferrable server algorithm for enhanced aperiodic responsiveness in hard real-time environments. IEEE Trans Comput 44(1):73–91
32. Tejera D, Tolosa R, de Miguel MA, Alonso A (2005) Two alternative rmi models for real-time distributed applications. In: ISORC, pp 390–397
33. Tejera D, Alonso A, de Miguel MA (2007) Predictable serialization in java. In: IEEE International Symposium on Object/Component/Service-Oriented Real-Time Distributed Computing (ISORC), pp 102–109
34. Tejera D, Alonso A, de Miguel MA (2007) Rmi-hrt: remote method invocation – hard real time. In: International Workshop on Java Technologies for Real-time and Embedded Systems (JTRES), pp 113–120
35. Tolosa R, Mayo JP, de Miguel MA, Higuera-Toledano MT, Alonso A (2003) Container model based on rtsj services. In: On the Move to Meaningful Internet Systems, pp 385–396

Chapter 8
The ASSL Formalism for Real-Time Autonomic Systems

Emil Vassev and Mike Hinchey

Abstract To increase the overall reliability, modern computerization of complex systems such as spacecraft, airplanes, and automobiles strives to minimize the human control via self-adaptation. In such systems, special autonomic components are designed to handle performance problems via monitoring and automatic decision making. Successful development of autonomic systems requires new software development approaches. ASSL (Autonomic System Specification Language) is a framework dedicated to the development of autonomic systems whereby developers are helped with problem formation, component and system design, analysis, evaluation, and implementation. The bottom line is a special multi-tier specification mechanism exposing a rich set of constructs allowing a system to be modeled by emphasizing different key aspects, but centering the model around special self-management policies. This entry presents in detail the aforementioned mechanism together with the underlying semantics. As a case study, we also present an ASSL specification model for autonomic components of an autonomous NASA space exploration system.

8.1 Introduction

Complexity is widely recognized as one of the biggest challenges information technology faces today. To respond to this threat, initiatives, such as Autonomic Computing (AC) [5, 6, 9, 10], have been started to deal with complexity in computerized systems. AC is a rapidly growing field that promises a new approach to developing large-scale complex systems capable of self-management. The phrase "autonomic computing" came into the popular consciousness at the AGENDA 2001

E. Vassev (✉) • M. Hinchey
Lero–the Irish Software Engineering Research Centre, University of Limerick, Limerick, Ireland
e-mail: emil.vassev@lero.ie; mike.hinchey@lero.ie

M.T. Higuera-Toledano et al. (eds.), *Self-Organization in Embedded Real-Time Systems*, 151
DOI 10.1007/978-1-4614-1969-3_8,

conference where Paul Horn from IBM presented the new computing paradigm by likening computer systems to the human Autonomic Nervous System [6]. The idea behind this is that software systems must manage themselves, as the human body does, or they risk being crushed under their own complexity. Many major software vendors, such as IBM, HP, Sun, and Microsoft have started research programs to create *self-managing computer systems*. However, their research efforts are mainly to make *individual components* of particular systems more self-managing rather than providing a complete solution to the problem of autonomic system development.

A typical, *autonomic component* is designed to perform self-adaptive behavior exhibited via dedicated functions, often with real-time constraints. In such components, both hardware and software are embedded as parts of a complete device or a bigger system, e.g., spacecraft. In addition, they often have long-life and 24x7 operational requirements where time constraints may trigger self-adaptation. The exhibited self-adaptive behavior makes such embedded systems more resilient and increases their overall reliability and quality. This and other advantages such as overall cost reduction due to maintenance minimization make AC very attractive to major industry areas such as automotive, production technique, avionics and space. For example, to increase the survivability of the remote missions, NASA is exploiting principles and techniques that help such systems become more resilient through self-management and automatic adaptation. By adhering to the principles of AC [14] contemporary spacecraft systems implement vital features for unmanned missions, such as *self-configuration*, *self-healing*, *self-optimization*, and *self-protection*.

This entry presents the formalism of the Autonomic System Specification Language (ASSL) [21], a dedicated to AC formal tool that emerges as a formal approach to developing autonomic systems. Providing both a formal notation and tools that support modeling and specification, validation and code generation of autonomic systems, ASSL has been successfully used in a variety of projects targeting functional prototypes of autonomous NASA space exploration missions [24, 27], autonomic pattern-recognition systems [25], home-automation sensor networks [28], etc. Note that a good understanding of the ASSL formalism and mastering the same were of major importance for the success of these endeavors. This entry gives an overview of the ASSL operational semantics and through formal semantics definitions it presents the operational behavior of the ASSL-specified self-healing model for space exploration missions [27]. This approach helps ASSL developers conceive an explicit understanding of the ASSL formalism.

The rest of this entry is organized as follows. In Sect. 8.2, we present different formalisms for autonomic systems. Section 8.3 describes the ASSL specification model. In Sect. 8.4, we present ASSL constructs for real-time properties. Section 8.5 presents the ASSL operational semantics, which is used in Sect. 8.6 to enlighten an ASSL specification of NASA space exploration missions. Section 8.6 also presents some test results and finally, Sect. 8.7 concludes the entry with summary remarks.

8.2 Formalism for Autonomic Systems

Conceptually, any formalism aims to assist the development of computer systems by providing formal notations that can be used to specify desirable system concepts (e.g. functionality). Usually, formal notations help developers precisely describe with the logical underpinning of mathematics features of the system under consideration at a higher level of abstraction than the one provided by implementation. However, a requirement is that developers should be able to move in a logical way from a formal specification of a system to implementation.

8.2.1 Formal Approaches to AC

Autonomic systems are special computer systems that emphasize *self-management* through *ontext awareness* and *self-awareness* [5, 6, 9, 10, 14]. Therefore, an AC formalism should not only provide a means of description of system behavior but also should tackle the vital for autonomic systems self-management and awareness issues. Moreover, an AC formalism should provide a well-defined semantics that makes the AC specifications a base from which developers may design, implement, and verify autonomic systems.

Formalisms dedicated to AC have been targeted by a variety of industrial and university projects. IBM Research developed a framework called Policy Management for Autonomic Computing (PMAC) [7, 8]. The PMAC formalism emphasizes the specification of self-management policies encompassing the scope under which those policies are applicable. A PMAC policy specification includes: 1) conditions to which a policy is in conformance (or not); 2) a set of resulting actions; 3) goals; and 4) decisions that need to be taken.

The so-called Specification and Description Language (SDL) is an object-oriented, formal language defined by the International Telecommunications Union—Telecommunications Standardization Sector (ITU-T) [19]. SDL is dedicated to real-time systems, distributed systems, and generic event-driven systems. The basic theoretical model of an SDL system consists of a set of extended finite state machines, running in parallel and communicating via discrete signals, thus making SDL suitable for the specification of self-management behavior.

Cheng et al. talk in [3] about a specification language for self-adaptation based on the ontology from system administration tasks and built over the underlying formalism of *utility theory* [17]. In this formalism, special self-adaptation actions are described as architectural operators, which are provided by the architectural style of the target system. A script of actions corresponds to a sequence of architectural operators. This sequence forms the so-called *adaptation tactic* defined in three parts: (1) the conditions of applicability; (2) a sequence of actions; and (3) a set of intended effects after execution. The definition of a tactic is similar to the "design by contract" interface definition [13].

Another formalism for ASs is provided by the so-called *chemical programming* (represented by the Gamma Formalism [2]) which uses the chemical reaction metaphor to express the coordination of computations. The Gama Formalism describes computation in terms of chemical reactions (described as rules) in solutions (described as multisets of elements). When applied to AS specification, the Gama Formalism captures the intuition of a collection of cooperative components that evolve freely according to some predefined constraints (rules). System self-management arises as a result of interactions between components, in the same way as "intelligence" emerges from cooperation in colonies of biological agents.

In [1], Andrei and Kirchner present a biologically inspired formalism for AC called Higher-Order Graph Calculus (HOGC). This approach extends the Gama Formalism with high-level features by considering a graph structure for the molecules and permitting control on computations to combine rule applications. HOGC borrows various concepts from graph theory, in particular from *graph transformations* [4], and use representations for graphs that have been already intensively formalized.

8.2.2 The ASSL Formalism

ASSL is a *declarative* specification language for autonomic systems with well-defined semantics. It implements modern programming language concepts and constructs like inheritance, modularity, type system, and high abstract expressiveness. Being a formal language designed explicitly for specifying autonomic systems (ASs) ASSL copes well with many of the AS aspects [6, 9, 10, 14]. Moreover, specifications written in ASSL present a view of the system under consideration, where specification and design are intertwined. Conceptually, ASSL is defined through formalization tiers (see Sect. 8.3). Over these tiers, ASSL provides a *multi-tier specification model* that is designed to be scalable and exposes a judicious selection and configuration of infrastructure elements and mechanisms needed by an AS. In order to determine the level of ASSL formalism, we investigated in the vast field of formal specification languages. Srivas and Miller in [18] refer to *constructive* versus *descriptive* style of specification (also known as *model-oriented* versus *property-oriented*). The constructive or model-oriented style is typically associated with the use of definitions, whereas the descriptive or property-oriented style is generally associated with the use of axioms [15]. ASSL benefits from both styles, by using a property-oriented axiomatization as a top-level specification style and introducing a suitable number of specification layers with increasingly detailed model-oriented descriptions. As a formal language, ASSL defines a neutral, implementation-independent representation for ASs. Similar to many formal notations, ASSL enriches the underlying logic with modern programming concepts and constructs thereby increasing the expressiveness of the formal language while retaining the precise semantics of the underlying logic. For example, the ASSL formalism for self-management policies (see Sect. 8.3.1.1) is based on *event calculus* [12], whose formalism is enriched to fit in the ASSL mechanism for specifying self-management policies [21].

To the best of our knowledge, the ASSL formalism is currently the only complete solution to the problem of AS specification. Although other solutions do exist, they emphasize individual AC aspects (e.g. self-management policies), which is far from what ASSL is proposing with its reach multi-tier specification model. Moreover, the ASSL framework together with the powerful formalism provides mature tools that allow ASSL specifications to be edited and formally validated. Finally, an operational Java application may be generated from any valid ASSL specification.

8.3 ASSL Specification Model

The ASSL formal notation is based on a specification model exposed over hierarchically organized *formalization tiers* [21]. This specification model provides both infrastructure elements and mechanisms needed by an AS. ASSL defines ASs with special *self-managing policies*, *interaction protocols* (IPs), and *autonomic elements* (AEs), where the ASSL tiers and their sub-tiers describe different aspects of the AS under consideration. Figure 8.1 presents the ASSL specification model. As shown, it decomposes an AS in two directions—(1) into levels of functional abstraction; and

```
I. Autonomic System (AS)
   * AS Service-level Objectives
   * AS Self-managing Policies
   * AS Architecture
   * AS Actions
   * AS Events
   * AS Metrics
II. AS Interaction Protocol (ASIP)
   * AS Messages
   * AS Communication Channels
   * AS Communication Functions
III. Autonomic Element (AE)
   * AE Service-level Objectives
   * AE Self-managing Policies
   * AE Friends
   * AE Interaction Protocol (AEIP)
     - AE Messages
     - AE Communication Channels
     - AE Communication Functions
     - AE Managed Elements
   * AE Recovery Protocol
   * AE Behavior Models
   * AE Outcomes
   * AE Actions
   * AE Events
   * AE Metrics
```

Fig. 8.1 ASSL multi-tier specification model

(2) into functionally related sub-tiers. The first decomposition presents the system from three different perspectives [21]:

1. *AS tier*—presents a general and global AS perspective, where we define the general autonomic system rules in terms of *service-level objectives (SLO)* and *self-management policies*, *architecture topology* and *global actions*, *events* and *metrics* applied in these rules.
2. *AS Interaction Protocol (ASIP) tier*—forms a communication protocol perspective, where we define the means of communication between AEs. An ASIP is composed of *channels*, *communication functions*, and *messages*.
3. *AE tier*—forms a unit-level perspective, where we define interacting sets of individual AEs with their own behavior. This tier is composed of AE rules (*SLO* and *self-management policies*), an *AE interaction protocol* (AEIP), *AE friends* (a list of AEs forming a circle of trust), *recovery protocols*, special *behavior models* and *outcomes*, *AE actions*, *AE events*, and *AE metrics*.

The second decomposition presents the major tiers AS, ASIP, and AE as composed of functionally related sub-tiers, where new AS properties emerge at each sub-tier. This allows for different approaches to AS specification. For example, we may start with a global perspective of the system by specifying the AS service-level objectives and self-management policies and by digging down to find the needed metrics at the very detail level of AE sub-tiers. Alternatively, we may start working at the detail level of AE sub-tiers and build our AS bottom-up. Finally, we can work on both abstract and detail level sides by constantly synchronizing their specification.

8.3.1 ASSL Tiers

The AS Tier specifies an AS in terms of *service-level objectives* (AS SLOs), *self-management policies*, *architecture topology*, *actions*, *events*, and *metrics* (see Fig. 8.1). The AS SLOs are a high-level form of behavioral specification that help developers establish system objectives such as performance. The self-management policies could be any of (but not restricted to) the four so-called self-CHOP policies defined by the AC IBM blueprint [9]: *self-configuring*, *self-healing*, *self-optimizing*, and *self-protecting*. These policies are driven by events and trigger the execution of actions driving an AS in critical situations. The metrics constitute a set of parameters and observables controllable by an AS. At the ASIP Tier, the ASSL framework helps developers specify an AS-level interaction protocol as a public communication interface, expressed with special *communication channels*, *communication functions*, and *communication messages*. At the AE Tier, the ASSL formal model exposes specification constructs for the specification of the system's AEs. Note that AEs are considered to be analogous to software agents able to manage their own behavior and their relationships with other AEs.

Note that ASSL targets only the AC features of a system and helps developers clearly distinguish the *AC features* from the *system-service features*. This is possible, because with ASSL we model and generate special *AC wrappers* in the form of ASs that embed the components of non-AC systems [21]. The latter are considered as *managed elements* controlled by the AS in question. Conceptually, a managed element can be any software or hardware system (or sub-system) providing services. Managed elements are specified per AE (see Fig. 8.1) where the emphasis is on the interface needed to control a managed element. It is important also to mention that the ASSL tiers and sub-tiers are intended to specify different aspects of an AS, but it is not necessary to employ all of them in order to model such a system. For a simple AS we need to specify: (1) the AEs providing self-managing behavior intended to control the managed elements associated with an AE; and (2) the communication interface. Here, self-management policies must be specified to provide such self-managing behavior at the level of AS (the AS Tier) and at the level of AE (AE Tier). The following subsections briefly present some of the ASSL sub-tiers.

8.3.1.1 Self-management Policies

The self-management behavior of an ASSL-developed AS is specified with the self-management policies. These policies are specified with special ASSL constructs termed *fluents* and *mappings* [21].

```
ASSELF_MANAGEMENT {
 SELF_HEALING {
  FLUENT inLosingSpacecraft {
   INITIATED_BY { EVENTS.spaceCraftLost }
   TERMINATED_BY { EVENTS.earthNotified }
  }
  MAPPING {
   CONDITIONS { inLosingSpacecraft }
   DO_ACTIONS { ACTIONS.notifyEarth }
  }
 }
} // ASSELF_MANAGEMENT
```

A fluent is a state where an AS enters with fluent-activating events and exits with fluent-terminating events. A mapping connects fluents with particular actions to be undertaken. Usually, an ASSL specification is built around self-management policies, thus making such a specification AC-driven. Self-management policies are driven by events and actions determined deterministically. The ASSL code above presents a sample self-healing policy specification.

8.3.1.2 ASSL Events

ASSL aims at event-driven autonomic behavior. Hence, to specify self-management policies, we need to specify appropriate events (see Sect. 8.3.1.1). Here, we rely on the reach set of event types exposed by ASSL [21]. To specify ASSL events, one may use logical expressions over SLOs, or may relate events with metrics (see the ASSL code below), other events, actions, time, and messages. Moreover, ASSL allows for the specification of special conditions that must be stated before an event is prompted.

```
EVENT newAsteroidDetected {
  ACTIVATION {
    CHANGED { AS.METRICS.numberOfAsteroids }
  }
}
```

8.3.1.3 ASSL Metrics

For an AS, one of the most important success factors is the ability to sense the environment and react to sensed events. Together with the rich set of events, ASSL imposes metrics as a means of determining dynamic information about external and internal points of interest. Although four different types of metric are allowed [21], the most important are the so-called resource metrics because those are intended to measure special managed element quantities. The following ASSL code shows the ASSL specification of a resource metric called *noObstacle*. As shown, this metric is related to the *obstacleSensor* managed element.

```
METRIC noObstacle {
  METRIC_TYPE { RESOURCE }
  METRIC_SOURCE { AEIP.MANAGED_ELEMENTS.obstacleSensor.isClean}
  THRESHOLD_CLASS { Boolean [ true ] }
}
```

8.3.1.4 Managed Elements

An AE typically controls *managed elements*. In an ASSL-developed AS, a managed element is specified with a set of special interface functions intended to provide control functionality over that managed element. Note that ASSL can specify and generate interfaces controlling a managed element (generated as a stub), but not the real implementation of these interfaces. This is just fine for prototyping, however when deploying an AS prototype the generated interfaces must be manually programmed to deal with the controlled system (or sub-system).

```
MANAGED_ELEMENT meReceptor {
  INTERFACE_FUNCTION reset {}
  INTERFACE_FUNCTION getRadiationLevel {
    PARAMETERS {DECIMAL xCoord; DECIMAL yCoord; DECIMAL zCoord}
    RETURNS { DECIMAL }
    TRIGGERS { AS.EVENTS.newRadiationLevel }
    ONERR_TRIGGERS { AS.EVENTS.cannotGetRadiationLevel }
  }
}
```

8.3.1.5 Interaction Protocols

The ASSL *interaction protocols* provide a means of communication interface expressed with *messages* that can be exchanged among AEs, *communication channels*, and *communication functions*. Thus, by specifying an ASSL interaction protocol we develop an embedded messaging system needed to connect the AEs of an AS. In a basic communication process ongoing in such a system, an AE relies on a communication function to receive a message over an *incoming* communication channel, changes its internal state and sends some new messages over an *outgoing* channel [21].

```
ASIP {
  MESSAGES { MESSAGE msgHello { SENDER { AES.ae2 } RECEIVER { AES.ae1 } } }
  CHANNELS {
    CHANNEL chnlIO { ACCETS { ANY } ACCESS { SEQUENTIAL } DIRECTION { INOUT}}
  }
  FUNCTIONS {
    FUNCTION sendHello {
      PARAMETERS { BOOLEAN hasSpeed; BOOLEAN hasDirection }
      DOES { MESSAGES.msgHello >> CHANNELS.chnlIO }
    }
  }
}
```

8.4 ASSL Features for Real-Time Properties

ASSL implies a number of important specification constructs and techniques that support the development of real-time systems.

8.4.1 Real-Time Events

In general, real-time systems are considered event-driven. ASSL exposes a rich set of techniques and constructs for specifying events (see Sect. 8.3.1.2, which makes the framework suitable for the specification and code generation of event-driven embedded systems. From the real-time development perspective, events are one of the most important aspects in ASSL. By its nature, an event is a means for high-priority system messaging. ASSL uses events to specify many of the ASSL tiers and sub-tiers, such as *fluents* of *self-management policies* (see Sect. 8.3.1.1, *actions*, etc. To specify events, one may use logical expressions over *service-level objectives* (SLO), *metrics*, other *events*, *messages*, etc. Here, in order to specify events, ASSL introduces the following clauses:

- *DEGRADED/NORMALIZED*—prompt an event when specified SLOs transit from normal to degraded state and from degraded to normal state, respectively;
- *RECEIVED/SENT*—to prompt an event when an ASSL message has been received or sent, respectively;
- *CHANGED*—to prompt an event when the value of a specific ASSL metric has been changed;
- *OCCURRED*—to prompt an event when another ASSL event has occurred;
- *ACTIV_TIME*—to prompt an event when a specific time has occurred;
- *PERIOD*—to prompt an event regularly on period basis;
- *DURATION*—to specify the event duration once it has been prompted.

In addition, ASSL introduces a *GUARDS* clause to event specification to define conditions that must be stated before an event can be prompted.

```
EVENTS {
  EVENT archiveTime {
    ACTIVATION { ACTIV_TIME { 12:00 AM } }
  }
  EVENT doArchive {
    GUARDS { METRICS.sufficientSpace.VALUE = true }
    ACTIVATION { OCCURRED { EVENTS.archiveTime } }
    DURATION { 1 hour }
  }
} // EVENTS
```

The ASSL code above shows a specification sample specifying two real-time events. The first one (named *archiveTime*) is a timed event that will be prompted at $12:00AM$ to notify the system that it is time to do archive. The second one (named *doArchive*) will be prompted by the first event, but only if the $sufficientSpace$ metric holds true,i.e., if there is sufficient storage space. Note that there are two real-time constraints in this sample: (1) exact activation time; and (2) task duration.

8.4.2 Real-Time Metrics

For an *embedded real-time system*, perhaps the most important success factor is the ability to sense both the hardware and the operational environment and to react to detected events. Together with the rich set of events, ASSL imposes *metrics* to gather information about external and internal stimuli, e.g., events raised by the hardware.

Recall that in ASSL, metrics are control parameters and observables that an AS can control and/or monitor (see Sect. 8.3.1.3). Four different types of metrics assist in the specification of real-time properties:

- *Resource metrics*—measure managed resource quantities;
- *Quality metrics*—measure system qualities such as performance, response time, etc.;
- *Scalar metrics*—monitor predefined dynamic AS variables;
- *Composite metrics*—a function of other metrics.

For the development of embedded real-time system the most important are *resource metrics* and *quality metrics* when associated with real-time constraints (e.g., response time). Note that the *managed resource* (see Sect. 8.3.1.4) could be the hardware controlled by the embedded system. In such a case, metrics are specified with a metric source that links the embedded AS with a hardware parameter the metric is going to measure.

```
METRICS { // increments when a failed node has been discovered
  METRIC numberOfFailedNodes {
    METRIC_TYPE { RESOURCE }
    METRIC_SOURCE {
      AEIP.MANAGED_ELEMENTS.STAGE_ME.countFailedNodes }
    DESCRIPTION { ''counts failed nodes'' }
    VALUE { 0 }
    THRESHOLD_CLASS { integer [0] } // valid only when holds 0
  }
}
```

In the sample above, the metric *numberOfFailedNodes* is updated via a special interface function called *countFailedNodes* and specified at the *STAGE_ME* managed element. The latter represents the controlled managed resource. Moreover, metrics are specified with a special range of acceptable values expressed via a special ASSL construct called *threshold class*. In general, a threshold class determines rules for *valid* and *invalid* metric values. Note that metrics are evaluated by ASSL as valid and invalid based on their metric value and can prompt events when a new value has been detected. Thus, if a measured value does not fit into the metric threshold class, it is counted as undesirable behavior that should be carried by the system. This mechanism is very useful, because we can specify metrics prompting events when a real-time system's deadline cannot be met and the system must switch to an alternative execution path. For example, the sample above specifies a metric, which is valid only when *zero*(0) holds.

8.4.3 Managed Resource

An AE typically controls a *managed resource* specified in ASSL as *managed elements* (see Sect. 8.3.1.4). A managed element is generally a functional unit, a hardware or software system that provides certain services. In an embedded AS, a managed element may represent the controlled piece of hardware. An AE monitors and interacts with its managed elements via the *interface functions* (see Sect. 8.3.1.4), which provide an abstraction of the managed elements. Interface functions help to form a simple communication model for interacting with the managed elements. This model forms an extra layer at the AEIP (AE interaction protocol) (see Fig. 8.1 and Sect. 8.3.1.5). The AEIP tier is normally used to specify a private communication protocol used by an AE to communicate with:

- Trusted AEs;
- Controlled managed elements.

```
AEIP {
  MANAGED_ELEMENTS {
    MANAGED_ELEMENT STAGE_ME {
      INTERFACE_FUNCTION countFailedNodes { RETURNS { integer } }
      // runs the replica of a failed node
      INTERFACE_FUNCTION runNodeReplica { PARAMETERS { NetNode node }
        ONERR_TRIGGERS { EVENTS.nodeReplicaFailed }
      }
    }
  }
} // AEIP
```

For embedded AS, at this tier we should emphasize the specification of the managed element representing the controlled hardware. As shown by the sample above, with ASSL we specify a managed element as a Java-like interface, i.e., as a named collection of functions without implementation. The parameter types and the return type of those functions are ASSL-predefined or custom-defined types. The managed element interface functions can be called by the ASSL actions to control the managed elements. In addition, these can be associated with ASSL metrics (see Sect. 8.4.2) to retrieve information from the hardware.

ASSL specifies *managed element interface functions* with four non-mandatory clauses: *PARAMETERS*, *RETURNS*, *TRIGGERS*, and *ONERR_TRIGGERS*. Here, the *TRIGGERS* and *ONERR_TRIGGERS* clauses are used to specify events triggered by an interface function. For example, in the sample above the *runNodeReplica* interface function is specified to trigger a *nodeReplicaFailed* event in case of erroneous execution. Recall that events drive self-management policies, which allows for handling hardware-related events, and thus, incorporating an event-driven behavior into an real-time AS.

8.5 ASSL Notation and Semantics

ASSL is a declarative specification language for ASs with well-defined semantics [21]. The language provides a powerful formal notation that enriches the underlying logic with modern programming language concepts and constructs such as: *inheritance*, *modularity*, *type system*, and *abstract expressiveness*. As a formal language, ASSL defines a neutral representation (i.e., implementation-independent) for ASs described as a set of interacting AEs. The following is a generic meta-grammar in Extended Backus-Naur Form (BNF) [11] presenting the syntax rules for specifying ASSL tiers. Note that this meta-grammar is an abstraction of the ASSL grammar, which cannot be presented here due to the complex structure of the ASSL specification model (see Sect. 8.3), where each tier has its own syntax and semantic rules. The interested reader is advised to refer to [21] for the complete ASSL grammar expressed in BNF and for the semantics of the language.

```
GroupTier --> FINAL? ASSLGroupTierId { Tier+ }
Tier --> FINAL? ASSLTierId TierName? { Data* TierClause+ }
TierClause --> FINAL? ASSLClauseId ClauseName? { Data* }
Data --> TypeDecl* | VarDecl* | CllctnDecl* | Statement*
TypeDecl --> CustTypeIdentifier
VarDecl --> Type VarIdentifier
CllcntDecl --> Type CustCllctnIdentifier
Type --> CustType | PredefType
Statement --> Assign-Stmnt | Loop | If-Then-Else | Cllctn-Stmnt
Loop --> Foreach-Stmnt | DoWhile-Stmnt | WhileDo-Stmnt
```

As shown in the grammar above, an ASSL tier is syntactically specified with an ASSL *tier identifier*, an optional *tier name* and a *content block* bordered by curly braces. Moreover, we distinguish two syntactical types of tier: *single tiers* (Tier) and *group tiers* (GroupTier) where the latter comprise a set of single tiers. Each single tier has an optional *name* (TierName) and comprises a set of special *tier clauses* (TierClause) and optional *data* (Data). The latter is a set of *data declarations* and *statements*. Data declarations could be: (1) *type declarations*; (2) *variable declarations*; and (3) *collection declarations*. Statements could be: (1) *loop statements*; (2) *assignment statements*; (3) *if-then-else statements*; and (4) *collection statements*. Statements can comprise *Boolean* and *numeric expressions*. In addition, although not shown in the grammar above, note that identifiers participating in ASSL expressions are either simple, consisting of a single identifier, or qualified, consisting of a sequence of identifiers separated by "." tokens.

8.5.1 ASSL Operational Semantics

The formal evaluation of the operational behavior of ASSL specification models is a stepwise evaluation of the specified ASSL tiers, where the latter are evaluated as state transition models in which operations cause a current state to evolve to a new state [21]. Thus, if we use the convention for semantic function in which σ states for a current state and σ' states for a new state then the state evolution caused by an operation Op is denoted as $\sigma \xrightarrow{Op(x_1,x_2,\ldots,x_n)} \sigma'$ where the operation $Op(x_1,x_2,\ldots,x_n)$ is an abstraction of a transition operation performed by the framework that potentially takes n arguments. All the arguments are evaluated to their expression value first, and then the operation is performed. Here, Op is a transition operation of type O^{trans}(see the set definition below).

$O^{trans}\{DegradSLO, NormSLO, FluentIn,$
$FluentOut, ActionMap, Action, Function,$
$MsgRcvd, MsgSent, Event, EventOver, Metric,$
$ChangeStruct, CreateAE, ExtClass, RcvryProtocol,$
$BhvrModel, MngRsrcFunction, Outcome\}$

In addition, the operational semantics of the ASSL tiers introduces the notion of *tier environment* ρ presenting the *host tier* of the sub-tiers or clauses under evaluation. Thus, we write $\rho \vdash_\sigma$ to mean that ρ is evaluated in context σ and $\rho \vdash_\sigma e \rightarrow e'$ to mean that, in a given tier environment ρ (host tier for the expression e) one step of the evaluation of expression e in the context σ results in the expression e'. Here, the context σ is defined by the tier content, i.e., sub-tiers, tier clauses, etc. Note that the ASSL tiers may participate in expressions where they are presented by their *TierName*. For example, *AS/AE SLO*, *AS/AE policies*, *fluents*, *AS/AE events*, and *AS/AE metrics* can participate in *Boolean* expressions, where they are evaluated as **true** or **false** in the context of their host tier based on their performance.

The following subsections present two algorithms implemented by the ASSL framework for operational evaluation of ASSL actions and self-management policies.

8.5.2 Operational Evaluation of ASSL Actions

From operational semantics perspective, the AS/AE Action tier is the most important and the most complex ASSL tier. The following is a partial EBNF grammar presenting syntactically that tier.

```
Action-Decl  --> ACTION IMPL? Action-Name { Action-Decl-Seqnce }
Action-Decl-Seqnce --> Params-Decl? Returns-Decl? Guards-Decl?
      Ensures-Decl? Var-Decl-Seqnce? Does-Decl OnErr-Does-Decl?
      Trigs-Decl? OnErr-Trigs-Decl?
```

ASSL actions comprise the tier clauses: PARAMETERS {...}, RETURNS {...}, GUARDS {...}, ENSURES {...}, DOES {...}, ONERR_DOES {...}, TRIGGERS {...}, and ONERR_-TRIGGERS {...}. Note that only the DOES

{...} clause is mandatory. The operational evaluation of an ASSL action follows the following algorithm:

1. Process the action guards, if any (GUARDS {...} clause):
   - If the guards are held then perform the action.
   - Otherwise, deny the action.
2. Evaluate the variable declarations, if any.
3. Process the DOES {...} clause:
   - If a return statement is hit, then stop the action and return a result.
   - Else, process all the statements until the end of the DOES {...} clause.
4. If the DOES {...} clause is evaluated correctly, then evaluate the ENSURES {...} clause (in respect to the TRIGGERS {...} clause):
   - If the ENSURES {...} clause is held then trigger notification events via the TRIGGERS {...} clause and exit the action normally.
   - Else, process the ONERR_DOES {...} clause and trigger error events via the ONERR_TRIGGERS {...} clause.
5. If an error occurs while evaluating the action clauses, then stop the evaluation process and:
   - Process the ONERR_DOES {...} clause (similar to the evaluation of the DOES {...} clause), if any.
   - Trigger error events via the ONERR_TRIGGERS {...} clause, if any.

8.5.3 Operational Evaluation of ASSL Policies

ASSL specifies policies with *fluents* and *mappings* (see Sect. 8.3.1.1). Whereas the former are considered as specific policy conditions, the latter map these conditions to appropriate actions. A partial presentation of the fluent grammar is the following:

```
Fluent-Decl --> FLUENT Fluent-Name { Fluent-Inner-Decl }
Fluent-Name --> Bool-Identifier
Fluent-Inner-Decl --> Initiates-Sqnce Terminates-Sqnce
Fluent-Inner-Decl --> Initiates-Sqnce
Initiates-Sqnce --> INITIATED_BY { Event-Names }
Terminates-Sqnce --> TERMINATED_BY { Event-Names }

Map-Decl --> MAPPING { Mapping-Inner-Decl }
Mapping-Inner-Decl --> Condition-Sqnce Action-Sqnce
Condition-Sqnce --> CONDITIONS { Fluent-Names }
Action-Sqnce --> DO_ACTIONS { Action-Calls ; Action-Calls-Forall }
```

An ASSL policy is evaluated based on its fluents. The operational evaluation of a fluent follows the following algorithm: If an event has occurred in the system then:

1. Process the INITIATED_BY {...} clause to check if that event can initiate the policy fluent f and if so, initiate that fluent:
   - If the policy fluent f has been initiated then process only the policy MAPPING {...} clauses comprising the fluent f in their CONDITIONS {...} clause.
   - Evaluate the CONDITIONS {...} clause and if the stated conditions are held then evaluate the DO_ACTIONS {...} clause to perform the actions listed there.
2. Process the TERMINATED_BY {...} clause to check if that event can terminate the previously-initiated policy fluent f and if so, terminate it.

The semantic rules 1 through to 2 present the operational semantics that cope with the algorithm stated above. In these rules, each premise is a system transition operation (see Sect. 8.5.1) such as $Event(ev)$, $FluentIn(f,ev)$, $FluentOut(f,ev)$, and $ActionMap(f,a)$.

$$(1)\quad \frac{\sigma \xrightarrow{Event(ev)} \sigma'}{f \vdash_{\sigma'} INITIATED\_BY\{ev_1,\ldots,ev_n\} \xrightarrow{FluentIn(f,ev)} \sigma''}\; ev \in \{ev_1,\ldots,ev_n\}$$

$$(2)\quad \frac{\sigma \xrightarrow{FluentIn(f,ev)} \sigma' \; \sigma' \xrightarrow{Event(ev)} \sigma''}{f \vdash_{\sigma''} TERMINATED\_BY\{ev_1,\ldots,ev_n\} \xrightarrow{FluentOut(f,ev)} \sigma'''}\; ev \in \{ev_1,\ldots,ev_n\}$$

$$(3)\quad \frac{\sigma \xrightarrow{FluentIn(f,ev)} \sigma'}{map \vdash_{\sigma'} CONDITIONS\{f_1,\ldots,f_n\} \xrightarrow{ActionMap(f,a)} \sigma''}\; f \in \{f_1,\ldots,f_n\}$$

$$(4)\quad \frac{\sigma \xrightarrow{ActionMap(f,a)} \sigma'}{map \vdash_{\sigma'} DO\_ACTIONS\{a_1,\ldots,a_n\} \xrightarrow{\forall a \in \{a_1,\ldots,a_n\} \bullet Action(a)} \sigma''}\; a \in A^{\sigma}$$

Here, A^{σ} is the finite set of actions in the context σ. The first premise in rule 2 evaluates whether the fluent f is initiated, i.e., only initiated fluents can be terminated [21].

8.6 Case Study: ASSL Specifications for NASA ANTS

8.6.1 NASA ANTS

The Autonomous Nano-Technology Swarm (ANTS) concept sub-mission PAM (Pro-specting Asteroids Mission) is a novel approach to asteroid belt resource exploration that provides for extremely high autonomy, minimal communication requirements with Earth, and a set of very small explorers with a few consumables [20]. These explorers forming the swarm are pico-class, low-power, and

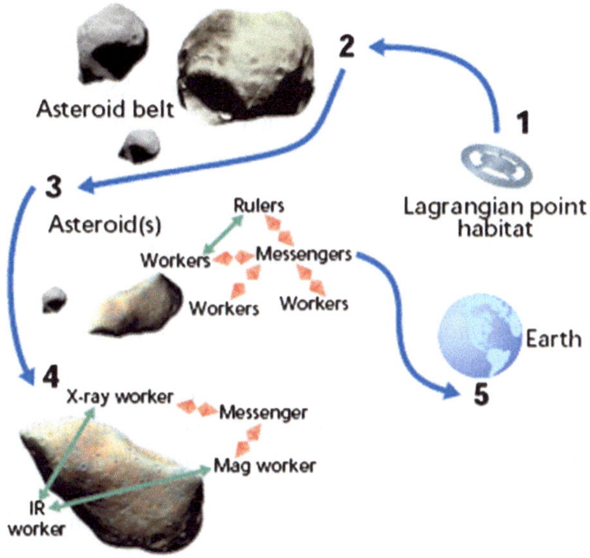

Fig. 8.2 ANTS mission concept

low-weight spacecraft, yet capable of operating as fully autonomous and adaptable units. The units in a swarm are able to interact with each other, thus helping them to self-organize based on the emergent behavior of the simple interactions. Figure 8.2 depicts the ANTS concept mission. A transport spacecraft launched from Earth toward the asteroid belt carries a laboratory that assembles the tiny spacecraft. Once it reaches a point in space, termed a Lagrangian, where gravitational forces on small bodies are balanced, the transport releases the assembled swarm, which will head for the asteroid belt. Each spacecraft is equipped with a solar sail for power; thus it relies primarily on power from the sun, using only tiny thrusters to navigate independently. Moreover, each spacecraft also has onboard computation, artificial intelligence, and heuristics systems for control at the individual and team levels. As Fig. 8.2 shows, there are three classes of spacecraft—*rulers*, *messengers*, and *workers*. Sub-swarms are formed to explore particular asteroids in an ant colony analogy. Hence, ANTS exhibits self-organization since there is no external force directing its behavior and no single spacecraft unit has a global view of the intended macroscopic behavior. The internal organization of a swarm depends on the global task to be performed and on the current environmental conditions. In general, a swarm consists of several sub-swarms, which are temporal groups organized to perform a particular task. Each swarm group has a group leader (*ruler*), one or more *messengers*, and a number of *workers* carrying a specialized instrument. The messengers are needed to connect the team members when they cannot connect directly, due to a long distance or a barrier.

8.6.2 Specifying ANTS with ASSL

In our endeavor to specify ANTS with ASSL, we emphasized modeling ANTS self-management policies such as self-configuring [27], self-healing [23], self-scheduling [26] and emergent self-adapting [22]. In addition, we proposed a specification model for the ANTS safety requirements [27]. To specify the ANTS safety requirements, we used the AS/AE SLO tier specification structures, and to specify the self-management policies we used ASSL tiers and clauses as following:

- *Self-management policies*—specify the self-management behavior of the system through a finite set of *fluents* and *mappings*.
- *Actions*—a finite set of actions that can be undertaken by ANTS in response to certain conditions, and according to the self-management policies.
- *Events*—a set of events that initiate fluents and are prompted by the actions according to the policies.
- *Metrics*—a set of metrics needed by the events and actions.

The following subsection presents the ASSL specification model for self-healing in ANTS together with a formal presentation of the operational behavior. Note that the specification presented here is partial, due to space limitations. The operational behavior of the self-healing specification is presented in a Structural Operational Semantics style [16]. Thus, we define semantics definitions formed by *inference rules*. An inference rule is presented as a set of premises deducting a conclusion, possibly under control of some condition.

8.6.3 Self-Healing

Figure 8.3 presents a partial specification of the self-healing policy for ANTS. In our approach, we assume that each worker sends, on a regular basis, heartbeat messages to the ruler [23]. The latter uses these messages to determine when a worker is not able to continue its operation, due to a crash or malfunction in its communication device or instrument. The specification shows only fluents and mappings forming the specification for the self-healing policy for an ANTS Worker. Here, the key features are:

- An *in Collision* fluent that takes place when the worker crashes into an asteroid or into another spacecraft, but it is still able to perform self-checking operations;
- An *in Instrument Broken* fluent that takes place when the self-checking operation reports that the instrument is not operational anymore;
- An *in Heartbeat Notification* fluent that is initiated on a regular basis by a timed event to send the heartbeat message to the ruler;
- A *check ANT Instrument* action that performs operational checking on the carried instrument.

```
AE ANT_Worker {
  AESELF_MANAGEMENT {
    SELF_HEALING {
      FLUENT inCollision {
        INITIATED_BY { EVENTS.collisionHappen } TERMINATED_BY
           { EVENTS.instrumentChecked } }
      FLUENT inInstrumentBroken {
        INITIATED_BY { EVENTS.instrumentBroken } TERMINATED_BY
           {  EVENTS.isMsgInstrumentBrokenSent } }
      FLUENT inHeartbeatNotification {
        INITIATED_BY { EVENTS.timeToSendHeartbeatMsg } TERMINATED_BY
           { EVENTS.isMsgHeartbeatSent } }
      MAPPING {//if collision then check if the instrument is still operational
        CONDITIONS { inCollision } DO_ACTIONS { ACTIONS.checkANTInstrument } }
      MAPPING { // if the instrument is broken then notify the group leader
        CONDITIONS { inInstrumentBroken } DO_ACTIONS
           { ACTIONS.notifyForBrokenInstrument } }
      MAPPING { // time to send a heartbeat message has come
        CONDITIONS { inHeartbeatNotification } DO_ACTIONS
           { ACTIONS.notifyForHeartbeat } }
    }
  }
  ...
    ACTIONS {
    ACTION  IMPL  checkInstrument { RETURNS {  BOOLEAN }  TRIGGERS
          { EVENTS.instrumentChecked } }
    ACTION checkANTInstrument {
      GUARDS { AESELF_MANAGEMENT.SELF_HEALING.inCollision }
      ENSURES { EVENTS.instrumentChecked }
      VARS { BOOLEAN canOperate }
      DOES {  canOperate = CALL  ACTIONS.checkInstrument }
      TRIGGERS { IF (not canOperate) THEN  EVENTS.instrumentBroken END }
      ONERR_TRIGGERS { IF (not canOperate) THEN  EVENTS.instrumentBroken END }
    }
    ...
  }
  ...
  EVENTS {
    EVENT collisionHappen {
      GUARDS { not METRICS.distanceToNearestObject }
      ACTIVATION { CHANGED { METRICS.distanceToNearestObject } } }
    EVENT timeToSendHeartbeatMsg { ACTIVATION { PERIOD { 1 min } } }
  }
  ...
  METRICS {
    METRIC distanceToNearestObject {
      METRIC_TYPE { RESOURCE }
      METRIC_SOURCE { AEIP.MANAGED_ELEMENTS.worker.getDistanceToNearestObject }
      THRESHOLD_CLASS { DECIMAL [0.001~) } }
  } // METRICS
} // ANT_Worker
```

Fig. 8.3 ASSL specification: self-healing

- A *distance To Nearest Object* metric that measures the distance to the nearest object in space (not presented here).
- A *collision Happened* event prompted by the *distance To Nearest Object* metric when the latter changes its value and the same does not satisfy the metric's threshold class.

8.6.3.1 Operational Behavior

A self-management policy is evaluated as "held" if the policy is not in either one of its specified fluents, and as "not held" if there is at least one initiated fluent for that policy (the policy is currently in that fluent) [21]. The SELF_HEALING policy (see Fig. 8.3) has three fluents: *inCollision*, *inInstrumentBroken*, and *inHeartbeatNotification*, i.e., the policy is evaluated as held when the policy is at least in one of these three fluents. Inference rules 5 through 35 enforce a definite strategy for evaluating the SELF_HEALING policy. The policy clauses (fluents and mappings) are evaluated in the context π of the SELF_HEALING policy, and the actions, events, and metrics are evaluated in the context of the *ANT_Worker* autonomic element (see Fig. 8.3). Inference rule 5 presents the operational evaluation of the *timeToSendHeartbeatMsg* timed event initiating the *inHeartbeatNotification* fluent (see rules 9 through 12). Thus,

$$(5)\quad \frac{a \vdash systemclock() \rightarrow t_{actv}}{ev \vdash_{\sigma} ACTIV\_TIME\{t_{actv}\} \xrightarrow{Event(timeToSendHeartbeatMsg)} \sigma'}$$

where *ev* is the tier environment exposed by the timed event, *systemclock*() is an abstract function returning the current time in the context σ, t_{actv} is the time when the timed event is specified to occur. Inference rules 6 through 8 present the operational evaluation of the *collisionHappened* event, which initiates the *inCollision* fluent (see rules 13–17). Thus,

$$(6)\quad \frac{\langle AE\rangle \xrightarrow{Metric(distanceToNearestObject)} \langle AE\rangle' \quad ev \vdash_{\sigma} distanceToNearestObject \rightarrow true}{ev \vdash_{\sigma} GUARDS\{distanceToNearestObject\} \rightarrow prompt(collisionHappened)}$$

$$(7)\quad \frac{\langle AE\rangle \xrightarrow{Metric(distanceToNearestObject)} \langle AE\rangle' \quad ev \vdash_{\sigma} distanceToNearestObject \rightarrow false}{ev \vdash_{\sigma} GUARDS\{distanceToNearestObject\} \rightarrow \neg prompt(collisionHappened)}$$

$$(8)\quad \frac{\langle AE\rangle \xrightarrow{Metric(distanceToNearestObject)} \langle AE\rangle'}{ev \vdash_{\sigma} CHANGED\{distanceToNearestObject\} \xrightarrow{Event(collisionHappened)} \sigma'}$$

where *ev* is the tier environment exposed by the *collisionHappen* event. In rules 6 and 7 we use the transition operation $Metric(distanceToNearestObject)$ to denote a state transition that occurs when the *distanceToNearestObject* metric changes its value, thus possibly prompting the *collisionHappened* event. Note that by operational semantic definition, an ASSL metric is evaluated as *Boolean* and is "true" only if the value it holds falls in the range determined by the metric's threshold class [21] (see THRESHOLD_CLASS in Fig. 8.3). Here, rules 6 and 7 evaluate the GUARDS {...} clause, which verifies whether that metric is still valid after changing its value.

Inference rules 9 through 12 present the operational evaluation of the *inHeartbeatNotification* fluent together with the MAPPING {...} clause mapping that fluent to the *notifyForHeartbeat* action. Thus,

$$(9)\quad \frac{\langle AE\rangle \xrightarrow{Event(timeToSendHeartbeatMsg)} \langle AE\rangle'}{f \vdash_{\sigma,\pi} INITIATED\_BY\{timeToSendHe...Msg\} \xrightarrow{FluentIn(f,timeToSendHe...Msg)} \sigma'\pi'}$$

$$(10)\quad \frac{\langle AE\rangle \xrightarrow{Event(msgHeartbeatSent)} \langle AE\rangle' \quad \pi \vdash_{\sigma} inHeartbeatNotification \rightarrow true}{f \vdash_{\sigma,\pi} TERMINATED\_BY\{msgHeartbeatSent\} \xrightarrow{FluentOut(f,msgHeartbeatSent)} \sigma'\pi'}$$

$$(11)\quad \frac{f \vdash_{\sigma,\pi} INITIATED\_BY\{timeToSendHe...Msg\} \xrightarrow{FluentIn(f,timeToSendHe..Msg)} \sigma'\pi'}{map \vdash_{\sigma',\pi'} CONDITIONS\{inHe...Notification\} \xrightarrow{ActionMap(f,notifyForHeartbeat)} \sigma''\pi''}$$

$$(12)\quad \frac{map \vdash_{\sigma',\pi'} CONDITIONS\{inHe...Notification\} \xrightarrow{ActionMap(f,notifyForHeartbeat)} \sigma''\pi''}{map \vdash_{\sigma'',\pi''} DO\_ACTIONS\{notifyForHeartbeat\} \xrightarrow{Action(notifyForHeartbeat)} \sigma'''\pi'''}$$

where f is the tier environment exposed by the *inHeartbeatNotification* fluent, π is the tier environment (and context) exposed by the SELF_HEALING policy, and *map* is the tier environment exposed by the MAPPING {...} clause (see Fig. 8.3).

Inference rules 13 through 17 present the operational evaluation of the *inCollision* fluent.

$$(13)\quad \frac{\langle AE\rangle \xrightarrow{Event(collisionHappened)} \langle AE\rangle'}{f \vdash_{\sigma,\pi} INITIATED\_BY\{collisionHappened\} \xrightarrow{FluentIn(f,collisionHappened)} \sigma'\pi'}$$

$$(14)\quad \frac{\langle AE\rangle \xrightarrow{Event(instrumentChecked)} \langle AE\rangle' \quad \pi \vdash_{\sigma} inCollision \rightarrow true}{f \vdash_{\sigma,\pi} TERMINATED\_BY\{instrumentChecked\} \xrightarrow{FluentOut(f,instrumentChecked)} \sigma'\pi'}$$

$$(15)\quad \frac{\langle AE\rangle \xrightarrow{Event(cannotCheckInstrument)} \langle AE\rangle' \quad \pi \vdash_{\sigma} inCollision \rightarrow true}{f \vdash_{\sigma,\pi} TERMINATED\_BY\{cannotCheckIns...nt\} \xrightarrow{FluentOut(f,cannotCheckIns...nt)} \sigma'\pi'}$$

$$(16)\quad \frac{f \vdash_{\sigma,\pi} INITIATED\_BY\{collisionHappened\} \xrightarrow{FluentIn(f,collisionHappened)} \sigma'\pi'}{map \vdash_{\sigma',\pi'} CONDITIONS\{inCollision\} \xrightarrow{ActionMap(f,checkANTInstrument)} \sigma''\pi''}$$

$$(17)\quad \frac{map \vdash_{\sigma',\pi'} CONDITIONS\{inCollision\} \xrightarrow{ActionMap(f,checkANTInstrument)} \sigma''\pi''}{map \vdash_{\sigma'',\pi''} DO\_ACTIONS\{checkANTInstrument\} \xrightarrow{Action(checkANTInstrument)} \sigma'''\pi'''}$$

Inference rules 18 through 21 present the operational evaluation of the *inInstrumentBroken* fluent (f is the tier environment exposed by that fluent).

$$(18)\quad \frac{\langle AE\rangle \xrightarrow{Event(instrumentBroken)} \langle AE\rangle'}{f \vdash_{\sigma,\pi} INITIATED\_BY\{instrumentBroken\} \xrightarrow{FluentIn(f,instrumentBroken)} \sigma'\pi'}$$

$$(19)\quad \frac{\langle AE\rangle \xrightarrow{Event(msgInstrumentBrokenSent)} \langle AE\rangle' \quad \pi \vdash_{\sigma} inInstrumentBroken \rightarrow true}{f \vdash_{\sigma,\pi} TERMINATED\_BY\{msgIn...BrokenSent\} \xrightarrow{FluentOut(f,msgIn...BrokenSent)} \sigma'\pi'}$$

$$(20)\quad \frac{f \vdash_{\sigma,\pi} INITIATED\_BY\{instrumentBroken\} \xrightarrow{FluentIn(f,instrumentBroken)} \sigma'\pi'}{\left(\frac{map \vdash_{\sigma',\pi'} CONDITIONS\{inInstrumentBroken\}}{\xrightarrow{ActionMap(f,notifyForBrokenInstrument)} \sigma''\pi''}\right)}$$

$$(21)\quad \frac{\left(\frac{map \vdash_{\sigma',\pi'} CONDITIONS\{inInstrumentBroken\}}{\xrightarrow{ActionMap(f,notifyForBrokenInstrument)} \sigma''\pi''}\right)}{map \vdash_{\sigma'',\pi''} DO\_ACTIONS\{notifyForBrknIn...t\} \xrightarrow{Action(notifyForBrknIn...t)} \sigma'''\pi'''}$$

Note that the *inInstrumentBroken* fluent is initiated by the *instrumentBroken* event (see Rule 18), which is triggered by the *checkANTInstrument* action (see Rule 27).

Inference rules 22 through 31 present the stepwise operational evaluation of the clauses of the *checkANTInstrument* action. Thus,

$$(22)\quad \frac{\langle AE\rangle \xrightarrow{Action(checkANTInstrument)} \langle AE\rangle' \quad a \vdash_{\sigma} collisionHappend \rightarrow true}{a \vdash_{\sigma} GUARDS\{collisionHappend\} \rightarrow perform(checkANTInstrument)}$$

$$(23)\quad \frac{\langle AE\rangle \xrightarrow{Action(checkANTInstrument)} \langle AE\rangle' \quad a\vdash_{\sigma} collisionHappend \rightarrow false}{a\vdash_{\sigma} GUARDS\{collisionHappend\} \rightarrow \neg perform(checkANTInstrument)}$$

$$(24)\quad \frac{\langle AE\rangle \xrightarrow{Action(checkANTInstrument)} \langle AE\rangle'}{a\vdash_{\sigma} DOES\{canOperate = CALL\ CheckInstrument\} \xrightarrow{Action(checkInstrument)} \sigma'}$$

$$(25)\quad \frac{\langle AE\rangle \xrightarrow{Action(checkANTInstrument)} \langle AE\rangle'}{a\vdash_{\sigma} DOES\{canOperate = CALL\ CheckInstrument\} \xrightarrow{\neg Action(checkInstrument)} \sigma'[err]}$$

$$(26)\quad \frac{\langle AE\rangle \xrightarrow{Action(checkANTInstrument)} \langle AE\rangle' \quad \left(a\vdash_{\sigma} DOES\{canOperate = CALL\ CheckInstrument\} \xrightarrow{Action(checkInstrument)} \sigma'\right)}{a\vdash_{\sigma'} TRIGGERS\{instrumentChecked\} \xrightarrow{Event(instrumentChecked)} \sigma''}$$

$$(27)\quad \frac{\langle AE\rangle \xrightarrow{Action(checkANTInstrument)} \langle AE\rangle' \quad \left(a\vdash_{\sigma'} TRIGGERS\{instrumentChecked\} \xrightarrow{Event(instrumentChecked)} \sigma'' \quad a\vdash_{\sigma''} be \rightarrow true\right)}{a\vdash_{\sigma''} TRIGGERS\{IF\ be\ THEN\ instrumentBroken\ END\} \xrightarrow{Event(instrumentBroken)} \sigma'''}$$

$$(28)\quad \frac{\langle AE\rangle \xrightarrow{Action(checkANTInstrument)} \langle AE\rangle' \quad \left(a\vdash_{\sigma'} TRIGGERS\{instrumentChecked\} \xrightarrow{Event(instrumentChecked)} \sigma'' \quad a\vdash_{\sigma''} be \rightarrow false\right)}{a\vdash_{\sigma''} TRIGGERS\{IF\ be\ THEN\ instrumentBroken\ END\} \rightarrow \sigma''}$$

$$(29)\quad \frac{\langle AE\rangle \xrightarrow{Action(checkANTInstrument)} \langle AE\rangle' \quad a\vdash_{\sigma} instrumentChecked \rightarrow true}{a\vdash_{\sigma} ENSURES\{instrumentChecked\} \rightarrow \sigma}$$

$$(30)\quad \frac{\langle AE\rangle \xrightarrow{Action(checkANTInstrument)} \langle AE\rangle' \quad a\vdash_{\sigma} instrumentChecked \rightarrow false}{a\vdash_{\sigma} ENSURES\{instrumentChecked\} \rightarrow \sigma'[err]}$$

$$(31)\quad \frac{\langle AE\rangle \xrightarrow{Action(checkANTInstrument)} \langle AE\rangle'}{a\vdash_{\sigma} ONERR\_TRIGGERS\{cannotCheckIns...t\} \xrightarrow{Event(cannotCheckIns...t)} \sigma'} \quad Err^{a} \neq \emptyset$$

where a is the tier environment exposed by the *checkANTInstrument* action, *be* states for a Boolean expression (evaluated in a single step), and Err^a is the finite set of errors produced by that action in a single performance of the $Action(checkANT-Instrument)$ transition operation. In addition, we use the $perform(a)$ abstract function to denote continuation of the *checkANTInstrument* action.

Inference rules 32 through 33 and rules 34 through 35 present the operational evaluation of *notifyForHeartbeat* and *checkANTInstrument* actions respectively. Note, that (1) due to space limitations, the ASSL specification of these actions is not presented in Fig. 8.3; (2) we present only the evaluation of their DOES {...} and TRIGGERS {...} clauses.

$$(32)\quad \frac{\langle AE\rangle \xrightarrow{Action(notifyForHeartbeat)} \langle AE\rangle'}{a\vdash_{\sigma} DOES\{\ CALL\ sendHeartbeat\} \xrightarrow{Function(sendHeartbeat)} \sigma'}$$

$$(33)\quad \frac{\langle AE\rangle \xrightarrow{Action(notifyForHeartbeat)} \langle AE\rangle' \quad \left(a\vdash_{\sigma} DOES\{CALL\ sendHeartbeat\} \xrightarrow{Function(sendHeartbeat)} \sigma'\right)}{a\vdash_{\sigma'} TRIGGERS\{msgHeartbeatSent\} \xrightarrow{Event(msgHeartbeatSent)} \sigma''}$$

$$(34)\quad \frac{\langle AE\rangle \xrightarrow{Action(notifyForBrokenInstrument)} \langle AE\rangle'}{a\vdash_{\sigma} DOES\{\ CALL\ sendInstrumentBroken\} \xrightarrow{Function(sendInstrumentBroken)} \sigma'}$$

$$(35)\quad \frac{\langle AE\rangle \xrightarrow{Action(notifyForBrokenInstrument)} \langle AE\rangle' \quad \left(a\vdash_{\sigma} DOES\{\ CALL\ sendInstrumentBroken\} \xrightarrow{Function(sendInstrumentBroken)} \sigma'\right)}{a\vdash_{\sigma'} TRIGGERS\{msgInstrumentBrokenSent\} \xrightarrow{Event(msgInstrumentBrokenSent)} \sigma''}$$

8.6.3.2 Testing the Self-Healing Behavior

In this experiment, we experimented with the Java generated code for the ASSL self-healing specification for ANTS [23]. Note that by default, any Java application generated with the framework generates run-time log records that show important state-transition operations ongoing in the system at runtime. Thus, we can easily trace the behavior of the generated system by following the log records generated by the same. In this test, we generated the Java application for the ASSL self-healing specification model for ANTS, compiled the same with Java 1.6.0, and ran the compiled code. The application ran smoothly with no errors.

First, it started all system threads as it is shown in the following log records. Note that starting all system threads first is a standard running procedure for all Java application skeletons generated with the ASSL framework.

```
Log Records ''Starting System Threads''

1)  METRIC 'generatedbyassl... metrics.DISTANCETONEARESTOBJECT': started
2)  EVENT 'generatedbyassl... events.INSTRUMENTLOST': started
3)  EVENT 'generatedbyassl... MSGINSTRUMENTBROKENRECEIVED': started
4)  EVENT 'generatedbyassl... SPACECRAFTCHECKED': started
5)  EVENT 'generatedbyassl... TIMETORECEIVEHEARTBEATMSG': started
6)  EVENT 'generatedbyassl... INSTRUMENTOK': started
7)  EVENT 'generatedbyassl... MSGHEARTBEATRECEIVED': started
8)  EVENT 'generatedbyassl... RECONFIGURATIONDONE': started
9)  EVENT 'generatedbyassl... RECONFIGURATIONFAILED': started
10) EVENT 'generatedbyassl... COLLISIONHAPPEN': started
11) FLUENT 'generatedbyassl... INHEARTBEATNOTIFICATION': started
12) FLUENT 'generatedbyassl... INCOLLISION': started
13) FLUENT 'generatedbyassl... INTEAMRECONFIGURATION': started
14) FLUENT 'generatedbyassl... INCHECKINGWORKERINSTRUMENT': started
15) POLICY 'generatedbyassl... SELF_HEALING': started
16) AE 'generatedbyassl.as.aes.ANT_RULER': started
******************************************************
17) METRIC 'generatedbyassl... DISTANCETONEARESTOBJECT': started
18) EVENT 'generatedbyassl... ISMSGHEARTBEATSENT': started
19) EVENT 'generatedbyassl... INSTRUMENTCHECKED': started
20) EVENT 'generatedbyassl... ISMSGINSTRUMENTBROKENSENT': started
21) EVENT 'generatedbyassl... COLLISIONHAPPEN': started
22) EVENT 'generatedbyassl... INSTRUMENTBROKEN': started
23) EVENT 'generatedbyassl... TIMETOSENDHEARTBEATMSG': started
24) FLUENT 'generatedbyassl... INHEARTBEATNOTIFICATION': started
25) FLUENT 'generatedbyassl... ININSTRUMENTBROKEN': started
26) FLUENT 'generatedbyassl... INCOLLISION': started
27) POLICY 'generatedbyassl... SELF_HEALING': started
28) AE 'generatedbyassl.as.aes.ANT_WORKER': started
******************************************************
29) EVENT 'generatedbyassl.as.ants.events.SPACECRAFTLOST': started
30) EVENT 'generatedbyassl.as.ants.events.EARTHNOTIFIED': started
31) FLUENT 'generatedbyassl... INLOSINGSPACECRAFT': started
32) POLICY 'generatedbyassl... SELF_HEALING': started
33) AS 'generatedbyassl.as.ANTS': started
***************** AS STARTED SUCCESSFULLY ****************
```

Here, records 1 through to 16 show the ANT_RULER autonomic element startup, records 17 through to 28 show the ANT_WORKER autonomic element startup, and records 29 through to 33 show the last startup steps of the ANTS autonomic system. After starting up all the threads, the system ran in idle mode for 60 s, when the timed event *timeToSendHeartbeatMsg* occurred. This event is specified in the ANT_WORKER to run on a regular time basis every 60 sec (see Fig. 8.3). The occurrence of this event activated the self-healing mechanism as shown in the following log records.

As we see from the log records, the self-healing behavior correctly followed the specification model. Records 34 through to 38 show the initiation and termination of the INHEARTBEATNOTIFICATION fluent. This resulted in the execution of the NOTIFYFORHEARTBEAT action (see record 36) that sends a heartbeat message to ANT_RULER[1](see record 37). Records 39 through to 43 show how this message is handled by the ANT_RULER. Records 44 through to 47 show how the INCHECKINGWORKERINSTRUMENT fluent is handled by the system. This fluent is initiated by the MSGHEARTBEATRECEIVED event. Next the CHECKWORKERINSTRSTATUS action is performed (see record 45), which resulted into the INSTRUMENTOK event (see record 46). The latter terminated the INCHECKINGWORKERINSTRUMENT fluent (see record 47). Records 48 through to 66 show that the system continued repeating the steps shown in records 34 though to 47. This is because the policy-triggering events are periodic timed events and the system did not encounter any problems (that could possibly branch the program execution) while performing the executed actions.

```
Log Records ''Self-healing Behavior''

34) EVENT 'generatedbyassl... TIMETOSENDHEARTBEATMSG': has occurred
35) FLUENT 'generatedbyassl... INHEARTBEATNOTIFICATION': has been initiated
36) ACTION 'generatedbyassl... NOTIFYFORHEARTBEAT': has been performed
37) EVENT 'generatedbyassl... ISMSGHEARTBEATSENT': has occurred
38) FLUENT 'generatedbyassl... INHEARTBEATNOTIFICATION': has been terminated

39) EVENT 'generatedbyassl... TIMETORECEIVEHEARTBEATMSG': has occurred
40) FLUENT 'generatedbyassl... INHEARTBEATNOTIFICATION': has been initiated
41) ACTION 'generatedbyassl... CONFIRMHEARTBEAT': has been performed
42) EVENT 'generatedbyassl... MSGHEARTBEATRECEIVED': has occurred
43) FLUENT 'generatedbyassl... INHEARTBEATNOTIFICATION': has been terminated

44) FLUENT 'generatedbyassl... INCHECKINGWORKERINSTRUMENT': has been initiated
45) ACTION 'generatedbyassl... CHECKWORKERINSTRSTATUS': has been performed
46) EVENT 'generatedbyassl... INSTRUMENTOK': has occurred
47) FLUENT 'generatedbyassl... INCHECKINGWORKERINSTRUMENT': has been terminated

48) EVENT 'generatedbyassl... TIMETOSENDHEARTBEATMSG': has occurred
49) FLUENT 'generatedbyassl... INHEARTBEATNOTIFICATION': has been initiated
50) ACTION 'generatedbyassl... NOTIFYFORHEARTBEAT': has been performed
51) EVENT 'generatedbyassl... ISMSGHEARTBEATSENT': has occurred
52) FLUENT 'generatedbyassl... INHEARTBEATNOTIFICATION': has been terminated

53) EVENT 'generatedbyassl... TIMETORECEIVEHEARTBEATMSG': has occurred
54) FLUENT 'generatedbyassl... INHEARTBEATNOTIFICATION': has been initiated
55) EVENT 'generatedbyassl... TIMETOSENDHEARTBEATMSG': has occurred

56) FLUENT 'generatedbyassl... INHEARTBEATNOTIFICATION': has been initiated
57) ACTION 'generatedbyassl... CONFIRMHEARTBEAT': has been performed
58) ACTION 'generatedbyassl... NOTIFYFORHEARTBEAT': has been performed
59) EVENT 'generatedbyassl... MSGHEARTBEATRECEIVED': has occurred
60) FLUENT 'generatedbyassl... INHEARTBEATNOTIFICATION': has been terminated

61) FLUENT 'generatedbyassl... INCHECKINGWORKERINSTRUMENT': has been initiated

62) EVENT 'generatedbyassl... ISMSGHEARTBEATSENT': has occurred
63) FLUENT 'generatedbyassl... INHEARTBEATNOTIFICATION': has been terminated

64) ACTION 'generatedbyassl... CHECKWORKERINSTRSTATUS': has been performed
65) EVENT 'generatedbyassl... INSTRUMENTOK': has occurred
66) FLUENT 'generatedbyassl... INCHECKINGWORKERINSTRUMENT': has been terminated
```

[1]The ASSL specification of ANT_Ruler is not presented here. The interested reader is advised to refer to [23].

This experiment demonstrated that the generated code had correctly followed the specified self-healing policy by reacting to the occurring self-healing events and, thus, providing appropriate self-healing behavior.

8.7 Summary

This entry has presented the formalism of ASSL (Autonomic System Specification Language) in terms of notation and operational semantics. ASSL is a domain-specific formal approach providing both formalism and tool support that help developers implement autonomic systems. It has been successfully used to develop prototype models for a variety of systems incorporating AC features and proven to be a valuable approach to problem formation, modeling, verification and implementation of autonomic systems with real-time properties. With ASSL, the formal specifications are automatically verified for consistency flaws and the provided synthesis approach of automatic code generation, guarantees consistency between a specification and the corresponding implementation.

ASSL implies a complex multi-tier hierarchy of *specification constructs* categorized as ASSL *tiers*, *sub-tiers* and *clauses*. Both *structural* and *functional relationships* form the semantic relations between the ASSL specification constructs. Whereas the ASSL multi-tier specification model imposes the structural relationships between tiers, sub-tiers and clauses, the ASSL operational semantics forms the functional relationships of the same. Conceptually, the ASSL operational semantics is driven by special state-transition operations and tier states. The operational evaluation of ASSL specifications is a stepwise evaluation of the specified ASSL tiers, sub-tiers and clauses, which are evaluated as state transition models where state-transition operations cause a current state to evolve to a new one.

Specifying with ASSL requires a good understanding of the ASSL formalism. This entry tackles this problem by introducing the ASSL formalism from both structural and operational perspectives. In addition, to demonstrate the theoretical concepts and flavor of the ASSL formalism, a case study example has presented ASSL specifications and their operational evaluation.

In conclusion, it should be noted that ASSL provides the IT community with an extremely needed and powerful framework for development of autonomic systems. Overall, ASSL is sufficiently generic and adaptable to accommodate most of the AC development aspects.

Acknowledgements This work was supported by Science Foundation Ireland grant 03/CE2/I303_1 to Lero—the Irish Software Engineering Research Centre at University of Limerick, Ireland.

References

1. Andrei O, Kirchner H (2009) A higher-order graph calculus for autonomic computing. In: Graph theory, computational intelligence and thought: essays dedicated to Martin Charles Golumbic on the occasion of his 60th birthday, Lecture Notes in Computer Science, Springer, vol 5420, pp 15–26
2. Banatre JP, Fradet P, Radenac Y (2007) Programming self-organizing systems with the higher-order chemical language. Int J Unconvent Comput 3(3):161–177
3. Cheng SW, Garlan D, Schmerl B (2006) Architecture-based self-adaptation in the presence of multiple objectives. In: Proceedings of the 2006 international workshop on self-adaptation and self-managing systems (SEAMS2006). ACM, New York, pp 2–8
4. Corradini A, Montanari U, Rossi F, Ehrig H, Heckel R, Lowe M (1997) Algebraic approaches to graph transformation – part i: Basic concepts and double pushout approach. In: Rozenberg G (ed) Handbook of graph grammars and computing by graph transformations, World Scientific Pub Co Inc, vol 1, pp 163–246
5. Dobson S, Sterritt R, Nixon P, Hinchey M (2010) Fulfilling the vision of autonomic computing. IEEE Comput 43(1):35–41
6. Horn P (2001) Autonomic computing: IBM's perspective on the state of information technology. Technical report, IBM T. J. Watson Laboratory
7. IBM Corporation (2005) Tivoli software: Defining service-level objectives. Technical report, IBM Tivoli
8. IBM Corporation (2005) Policy management for autonomic computing - version 1.2. Technical report, IBM Tivoli
9. IBM Corporation (2006) An architectural blueprint for autonomic computing. Technical report, IBM Corporation
10. Kephart JO, Chess DM (2003) The vision of autonomic computing. IEEE Comput 36(1):41–50
11. Knuth DE (1964) Backus normal form vs. backus naur form. Comm ACM 7(12):735–773
12. Kowalsky R, Sergot M (1986) A logic-based calculus of events. New Generat Comput 4(1): 67–95
13. Leavens G, Cheon Y (2006) Design by contract with jml. Technial report, Formal Systems Laboratory (FSL) at UIUC
14. Murch R (2004) Autonomic computing: on demand series. IBM, Prentice Hall, Englewood Cliffs
15. NASA (1998) Formal methods specification and analysis guidebook for the verification of software and computer systems. I: Planning and technology insertion. NASA, Washington, DC
16. Plotkin GD (1981) A structural approach to operational semantics – report DAIMI FN-19. Technical report, Computer Science Department, Aarhus University, Aarhus, Denmark
17. Read D (2004) Utility theory from jeremy bentham to daniel kahneman—working paper no: Lseor 04-64. Technical report, Department of Operational Research, London School of Economics, London
18. Srivas M, Miller S (1995) Formal verification of the aamp5 microprocessor: A case study in the industrial use of formal methods. In: Proceedings of the workshop on industrial-strength formal specification techniques (WIFT '95), pp 2–16. IEEE Computer Society, Silver Spring
19. The International Engineering Consortium (none) Web ProForum Tutorials: Specification and Description Language (SDL). iec.org. http://www.iec.org
20. Truszkowski W, Hinchey M, Rash J, Rouff C (2004) Nasa's swarm missions: The challenge of building autonomous software. IT Profess 6(5):47–52
21. Vassev EI (2008) Towards a framework for specification and code generation of autonomic systems. Ph.D. thesis, Department of Computer Science and Software Engineering, Concordia University, Montreal
22. Vassev E, Hinchey M (2008) Assl specification of emergent self-adapting for nasa swarm-based exploration missions. In: Proceedings of the 2nd IEEE international conference on self-

adaptive and self-organizing systems workshops (SASOW 2008). IEEE Computer Society, Silver Spring, pp 13–18
23. Vassev E, Hinchey M (2009) Assl specification and code generation of self-healing behavior for nasa swarm-based systems. In: Proceedings of the 6th IEEE international workshop on engineering of autonomic and autonomous systems (EASe'09). IEEE Computer Society, Silver Spring, pp 77–86
24. Vassev E, Hinchey M (2009) Modeling the image-processing behavior of the nasa voyager mission with assl. In: Proceedings of the 3rd IEEE international conference on space mission challenges for information technology (SMC-IT'09). IEEE Computer Society, Silver Spring, pp 246–253
25. Vassev E, Mokhov S (2010) Towards autonomic specification of distributed marf with assl: Self-healing, In: Software Engineering Research, Management and Applications. Studies in Computational Intelligence, Springer, vol 296, pp 1–15
26. Vassev E, Hinchey M, Paquet J (2008) A self-scheduling model for nasa swarm-based exploration missions using assl. In: Proceedings of the 5th IEEE international workshop on engineering of autonomic and autonomous systems (EASe'08). IEEE Computer Society, Silver Spring, pp 54–64
27. Vassev E, Hinchey M, Paquet J (2008) Towards an assl specification model for NASA swarm-based exploration missions. In: Proceedings of the 23rd annual ACM symposium on applied computing (SAC 2008) – AC Track. ACM, New York, pp 1652–1657
28. Vassev E, Hinchey M, Nixon P (2010) Prototyping home automation wireless sensor networks with assl. In: Proceedings of the 7th International Conference on Autonomic Computing and Communications (ICAC2010). ACM, pp 71–72

Chapter 9
Organic Real-Time Middleware

Uwe Brinkschulte, Mathias Pacher, Alexander von Renteln, and Benjamin Betting

Abstract Acquiring Self-X properties like self-organization and self-healing is a main goal of Organic Computing—a new and very promising research area. Inspired by nature, Organic Computing research wants to learn and adopt from techniques and properties of nature. The artificial hormone system (AHS), motivated by the hormone system of mammals as a role model, was designed to map tasks in real-time on processing elements using artificial hormones. In previous publications, the idea of an organic middleware and first theoretical results were presented. In this contribution we evaluate the degree of order introduced by task allocation of the AHS, thus computing the so called emergence value. We also present ways of improving the performance and minimizing the memory consumption of an AHS implementation. In addition, we developed a dependable architecture featuring a generalized core and task concept for a mixed-signal multi-core System on Chip.

9.1 Introduction

Today's computational systems are growing increasingly complex. They are build from large numbers of heterogeneous processing elements with highly dynamic interaction. Middleware is a common layer in such distributed systems, managing the cooperation of tasks on the processing elements and hiding the distribution from the application. It is responsible for seamless task interaction on distributed hardware. All tasks are interconnected by the middleware layer and are able to operate beyond processing element boundaries as if they would reside on a single hardware platform. To handle the complexity of today's—furthermore tomorrow's—distributed systems, self-organization techniques are necessary. The idea to autonomously

U. Brinkschulte (✉) • M. Pacher • A. von Renteln • B. Betting
Goethe Universität, Robert Mayer-StraSSe 11-15, Frankfurt am Main, Germany
e-mail: brinks@es.cs.uni-frankfurt.de; pacher@es.cs.uni-frankfurt.de; renteln@es.cs.uni-frankfurt.de; betting@es.cs.uni-frankfurt.de

M.T. Higuera-Toledano et al. (eds.), *Self-Organization in Embedded Real-Time Systems*, DOI 10.1007/978-1-4614-1969-3_9,

achieve this desired behavior is introduced in several papers [13, 21, 30]. Such a system should be able to find a suitable initial configuration by itself, to adapt or optimize itself to changing environmental and internal conditions, to heal itself in case of system failures or to protect itself against attacks.

Middleware is well suited to realize self-X features like self-configuration, self-optimization, and self-healing. So we developed the concept of an organic middleware which is able to assign tasks to heterogeneous processors in a distributed system holding the above mentioned self-X features. We presented the concept of the artificial hormone system[1] (AHS) and evaluation results of an AHS simulator and an AHS middleware in various publications, see Sect. 9.2.

In this paper, we investigate three new topics concerning the AHS: Emergence of the AHS, minimizing memory usage for better embedded suitability, and changes towards a mixed signal processing SoC.

The paper is structured as follows: We present the State of the Art in Sect. 9.2. In Sect. 9.3 we explain our approach for decentralized task mapping on heterogeneous processing elements and the concept of the AHS. After the AHS introduction, we talk about emergence within the AHS. Section 9.4 gives with an overview, while Sect. 9.4.1 presents results of emergence measuring tests. In Sect. 9.5, we talk about the implementation of the AHS, problems we encountered, and how we solved them. Another important topic, adapting the AHS towards an mixed signal processing SoC, is discussed in Sect. 9.6. Here we show how we developed a dependable architecture featuring a generalized core and task concept for a mixed-signal multi-core System on Chip. Based on the hardware redundant mixed-signal multi-core structure, the AHS acts as an organic middleware to assign tasks to suitable cores in a decentralized way. To enable mixed signal processing, we extended the AHS from the digital to the analog domain. Furthermore, facing the overall development and validation of the MixedCoreSoC architecture, different sample digital cores have been assembled and placed on real hardware, using current FPGA technique. The summaries of the three main topics emergence, implementation, and mixed signal SoC are provided at the end of the corresponding sections. The paper ends with some final remarks and future work in Sect. 9.7.

[1]The term "artificial hormone system" was chosen, because our approach was highly inspired by the hormone system of higher mammals. There are several comparable properties between the hormone system in biology and our technical system. However, it has to be stated that the AHS is not a copy of the biological hormone system, but is rather inspired by nature and its strategies. In biology, hormones are chemical objects transmitted via chemical processes and reactions. In our approach, the messengers are bits and bytes transferred via communication links. As a result, the effects and principles of the biological hormone system and the AHS are similar; hence we dubbed the messengers in our approach "hormones" as well.

9.2 State of the Art

Self-organization has been a research focus for several years. Publications like [16] or [31] deal with basic principles of self-organizing systems, like e.g. emergent behavior, reproduction etc. Regarding self-organization in computer science, several projects and initiatives can be listed.

IBM's Autonomic Computing project [14, 17] deals with self-organization of IT servers in networks. Several so called self-X properties like self-optimization, self-stabilization, self-configuration, self-protection and self-healing have been postulated. To realize these properties, the MAPE cycle consisting of *M*onitor, *A*nalyze, *P*lan and *E*xecute has been defined. This MAPE cycle is executed in the background and in parallel to normal server activities similar to the autonomic nervous system.

The German *Organic Computing* Initiative has been founded in 2003. Its basic intention is to improve the controllability of complex embedded systems by using principles found in organic entities [26, 29]. Organization principles successful in biology are adapted to embedded computing systems. To deepen research on this topic, the priority program 1183 "Organic Computing" [9] of the *German Research Foundation (DFG)* had been established. Several projects within this priority program dealt with the increase of reliability for embedded systems. Some examples shall be mentioned in the following. A principle often used in Organic Computing to build reliable self-organizing systems is the "Observer/Controller Architecture" [25]. Here, the system under observation was expected to show emergent effects, e.g. traffic jams or density waves in the project "Organic Traffic Control" [22]. An observer is detecting such emergent effects while the controller tries to suppress unwanted emergent effects (e.g. jams) and to support positive emergent effects (e.g. fluent traffic). Learning classifier systems are a widely used technique to build such observer/controller systems. In the project "Organic Fault Tolerant Control Architecture for Robotic Applications" [20], the health status of mobile robots was permanently monitored by so-called Organic Control Units, which were attached to the Basic Control Units implementing the regular behavior. The Organic Control Units applied learning strategies, e.g. guarded online learning, to detect and counteract failures like defective legs by adapted control of the other legs. The project "Organic Self-organizing Bus-based Communication Systems" [32] researched a decentralized approach to self-organize communication bus parameters like sending rates, priorities, etc. Theoretical models based on gaming theory were applied. The project "Organic Computing Middleware for Ubiquitous Environments" [28] provided and researched a middleware architecture using a two level approach for self-configuration, self-optimization and self-healing. A fast reflex driven mechanism provided quick reactions to changes in the ubiquitous system, while a slower planning based algorithm was introduced for long-term goal driven modifications. The ASoC project [18] envisioned a self-organizing failure-resistant System on Chip architecture based on a MAPE cycle similar to IBM's Autonomic Computing.

The DoDOrg project [1] researched the use of bio-inspired principles to build a new, self-organizing robust processor architecture. As a basic concept, the AHS was used to assign software tasks to distributed processing nodes. The AHS was firstly presented in [2, 3].

Further publications like [23, 24] showed theoretical as well as first practical results. A basic proof-of-concept hormone simulator running on one computer was developed and first tests were presented. The feasibility and reliability of the AHS were examined and we were able to determine upper bounds for self-configuration as well as self-healing of the system. Publications like [4,5] dealt with the influence of different hormone levels and the effects they have towards the quality of the task mapping and the behavior of the AHS in general. Theoretical analysis brought us to establishing equations for upper and lower bounds of task mappings.

Current research deals with the notion of trust in organic computing systems [8] and resource-aware programming support [11].

On an international level, organic computing is followed by a task force of the IEEE Computational Intelligence Society (IEEE CIS ETTC OCTF) [15]. Also other international research programs have addressed self-organization aspects for computing systems, e.g. [7, 10].

9.3 The Artificial Hormone System Middleware

The aim of the AHS is to assign tasks to processors or processor cores, also called processing elements (PEs), in a self-organizing way. To achieve this, it uses three main types of hormones:

Eager value

This hormone type determines the suitability of a PE to execute a task. The higher the hormonal value the better the ability of the PE to execute the task.

Suppressor

This hormone type lowers the suitability of a task execution on a PE. Suppressors are subtracted from eager values. There exist several subtypes of suppressors, e.g., the *task suppressor* to prevent duplicate task allocation, the *monitoring suppressor* to indicate a deteriorating PE state and the *load suppressor* to indicate the current load of a PE caused by the executed tasks. While the task suppressor is broadcasted to all PEs in the system, the monitoring and load suppressors are only used locally to limit the number of executed tasks on a PE.

Accelerator

This hormone type favors the execution of a task on a PE. Accelerators are added to eager values. Like for suppressors there exist several subtypes of accelerators. The most important one is the *organ accelerator* used to cluster cooperating tasks in the neighborhood. This accelerator is multicasted to neighbor nodes to form so called 'virtual organs' of cooperating tasks. Another subtype is the *monitoring accelerator*, which is used locally to indicate improved PE capabilities.

More details on these subtypes of hormones are presented when needed because they are used for fine tuning of the AHS and do not contribute to its basic understanding.

We have to distinguish between received hormones and hormones to be sent and also between tasks and processors. Therefore, we use latin letters such as i as task indices and greek letters such as γ as processor indices. A hormone of any type denoted as $H^{i\gamma}$ with superscripted indices signifies that this hormone is dedicated to and will be received by PE γ and task T_i. For the operation of the hormone loop it is not necessary to know who is the sender of this hormone. A hormone of any type denoted as $H_{i\gamma}$ signifies that this hormone is sent by PE γ and task T_i to other PEs. The receiver of this hormone depends on the hormone type, as can be seen below.

The task assignment happens in the following way: Each PE periodically executes the hormone based control loop presented in Fig. 9.1. Each iteration consists of three stages.

Receive stage

PE γ receives the modified eager values $Em^{i\gamma}$, suppressors $S^{i\gamma}$ and accelerators $A^{i\gamma}$ for each task T_i from each PE in the network. The communication between the different PEs is depicted by the dashed lines.

Compute and decision stage

PE γ computes the modified eager values $Em_{i\gamma}$ for all of its tasks in the following way. The local static eager value $E_{i\gamma}$ indicates how suited γ is to execute task T_i. From this value, all suppressors $S^{i\gamma}$ received by task T_i are subtracted, and all accelerators received by task T_i are added:

$$Em_{i\gamma} = E_{i\gamma} - \sum S^{i\gamma} + \sum A^{i\gamma}$$

The modified eager value $Em_{i\gamma}$ of each task T_i is then waiting to be broadcasted to task T_i on the other PEs in the send stage.

To decide on the allocation of a task T_i a PE compares its own modified eager value $Em_{i\gamma}$ with the received modified eager values $Em^{i\gamma}$ (from all other PEs) for this

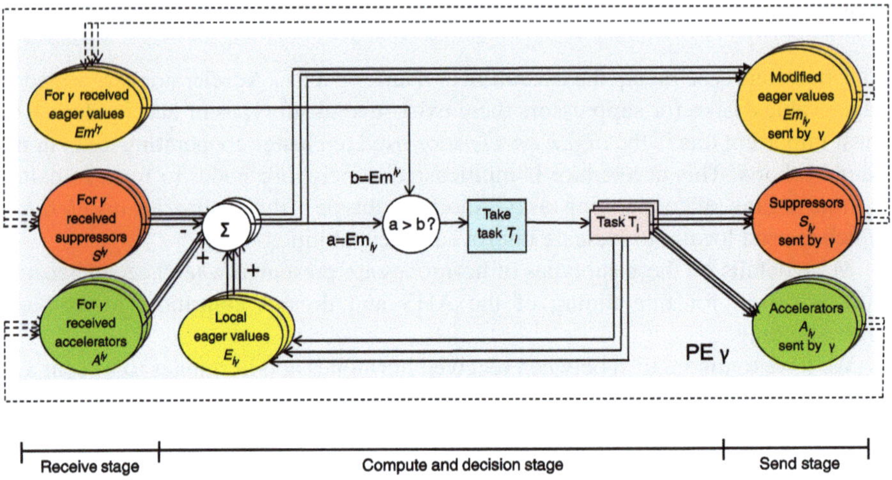

Fig. 9.1 Hormone based control loop

task. If $Em_{i\gamma} > Em^{i\gamma}$ is true for all received modified eager values $Em^{i\gamma}$ it decides to take the task. In case of equality, a second criterion e.g., the smallest position identifier of the PEs, is used to get an unambiguous decision. Otherwise another PE has the highest modified eager value for T_i and γ decides not to take it.

A PE selects the tasks for decision in a cyclic way i.e., each task will be selected in each mth iteration if m tasks have to be assigned. By selecting only one task at each iteration of the hormone loop the suppressors and accelerator can take effect. Otherwise the decision of taking a task would happen instantaneously and the hormones would have no effect.

Send stage

As already mentioned above, PE γ broadcasts the modified eager values $Em_{i\gamma}$ to each task T_i on the other PEs. The strength of these values depends on the results of the computation in the last phase.

If a task T_i is taken on PE γ, it also broadcasts a suppressors $S_{i\gamma}$ dedicated to the same task on all other PEs. On the one hand sending the suppressors indicates the PE has taken the tasks, and on the other hand it limits the number of allocations of the tasks.

Furthermore, the PE multicasts an accelerators $A_{i\gamma}$ to its neighbored PEs to attract tasks cooperating with task T_i to neighbored PEs thus forming clusters of tasks. Additionally, local suppressors to indicate the load of a PE and local accelerators to keep tasks from migrating too easily thus avoiding task oscillation are spread.
In this paper, we assume that the hormone based control loop and the sending and receiving of hormones perform perfectly on each PE. Detection and defense of attacks and failures are part of the future work, see Sect. 9.7.

Our approach is completely decentralized, each PE is responsible for its own tasks and the communication to other PEs is realized by a unified hormone concept. The AHS offers the following self-X-properties:

- The approach is self-organizing because no external influence controls the task allocation.
- It is self-configuring, an initial task allocation is found by exchanging hormones. The self-configuration is finished as soon as all modified eager values become zero, meaning no more tasks can to be taken.
- The self-optimization is done by re-offering tasks for allocation. The point of time for such an offer is determined by the task or by the PE. It can be done periodically or at a point of time when the task or the PE is idle.
  In this context, it is simple to handle the entrance of a new task in the system: At first, all PEs have to be informed about their hormone values for the new task. Then, the task is allocated as described for self-optimization.
- The approach is self-healing: In case of a task or PE failure all related hormones are sent no further. This results in an automatic reassignment of the task to the same PE (if it is still active) or to another PE.

In addition, the self-configuration, self-optimization and self-healing are real-time capable. Tight upper time bounds are given, these are presented in detail in [2,3,23]. We explain the main results here: Time is measured in (hormone) cycles. One cycle is the waiting time needed by a PE between sending the hormones and making the decision on executing a task.

Let us assume m tasks have to be assigned to the system ($m \in \mathbb{N}$). The worst-case time t_C that is used for complete self-configuring is:

$$t_C = 2m - 1 \text{ cycles} \tag{9.1}$$

For the worst-case self-optimization time t_O holds:

$$t_O = t_C$$

This means that all tasks are reallocated. In case only one task has to be reallocated, the time for self-optimization is 1 cycle.

The worst-case time t_H for self-healing is the same as for self-optimization plus a cycles needed to recognize a task failure ($a \in \mathbb{N}$):

$$t_H = t_O + a \text{ cycles}$$

The time a to recognize a task failure depends on the lifetime of a hormone, more precisely on the suppressor's hormone lifetime. When receiving a hormone the PE will mark it with a time stamp which is updated each time a new hormone arrives for this task. If no new hormone arrives the hormone expires and will be deleted after a hormone cycles (e.g., four hormone cycles). This is when a task failure is recognized.

9.4 Emergence

In this section, we consider the ability of the AHS of task clustering to organs regarding *emergence*. Emergence is a term used for a huge variety of effects in nature. The development of ant trails is a good example to demonstrate an emergent system, see Fig. 9.2.

When ants are searching for food they often find different trails from the anthill to a food source. This is shown by trail 1 and trail 2 in the figure. However, the ants are using only the shorter trail 1 to the food source after a while. This is more effective than using other trails like trail 2 regarding the transportation rate of amount of food per time.

The emergent effect in this example is that the ants are always able to find the shortest trail to the food source although none of them knows the trail. In fact, they manage to find the shortest trail by releasing pheromones. The concentration of pheromones will always be higher on a shorter trail than on all longer trails because the pheromones evaporate by and by. The ants are attracted by the pheromones thus following the trail with the highest pheromone concentration. This leads to the effect of ants always finding the shortest trail to the food source.

This and other examples lead to an abstract characterization of emergent systems: The often complex behavior of an emergent system results of simple behavior of its components. These components use simple rules for their actions and have no idea of the resulting system behavior.

There are many definitions of emergence and emergent systems similar to the above mentioned example. Most of them define emergence in a qualitative way. There is only a small number of definitions of emergence which deal with quantitative measures like [6, 12].

In [19] the authors propose a useful way of measuring the emergence of a system based on entropy. The entropy H on an alphabet Z ($|Z| = k$) is defined (see [27]) as

$$H := -\sum_{i=1}^{k} p_i \log_2(p_i)$$

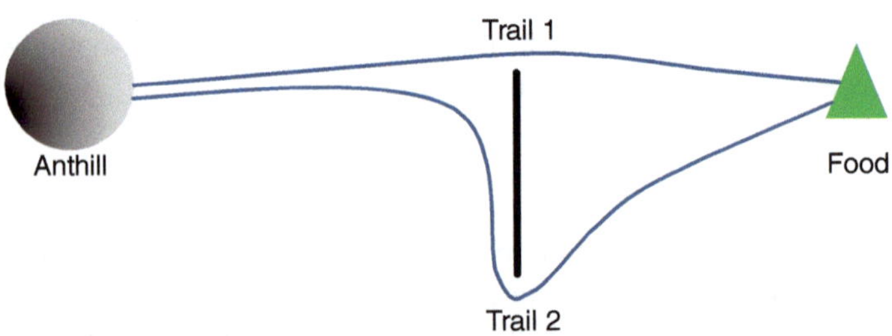

Fig. 9.2 Development of ant trails

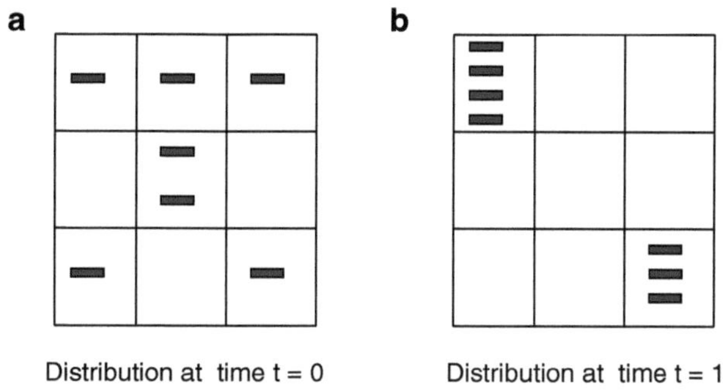

Fig. 9.3 Distribution of several items at different points in time

where p_i is the probability of the occurrence of an letter $z_i \in Z$ and $\sum_{i=1}^{k} p_i = 1$ holds. It is easy to compute that $0 \leq H \leq H_{\max} = \log_2 k$ is true. The (absolute) value of the emergence M of a system is then defined as difference of the maximal entropy of a system $H_{\max}$ and the current entropy of the system at point in time t:

$$M(t) := H_{\max} - H_t. \tag{9.2}$$

This definition makes sense: As higher the current entropy of a system is (there is much disorder in the system) as lower the value of the absolute emergence is. Vice versa, as lower the current entropy of the system is (the system is ordered in some way) as higher the value of the absolute emergence is.

A system is called emergent if $M > 0$ holds and the system is self-organizing.

Figure 9.3 shows an example: We assume to distribute items on a 3×3-array. The maximal entropy is $H_{\max} = \log_2 9 \approx 3.17$. The entropy at time $t = 0$ computes to $H_0 = 5 \cdot \frac{1}{7} \cdot \log_2(\frac{1}{7}) + 1 \cdot \frac{2}{7} \cdot \log_2(\frac{2}{7}) \approx 2.52$ and the entropy at time $t = 1$ computes to $H_1 = \frac{4}{7} \cdot \log_2(\frac{4}{7}) + \frac{3}{7} \cdot \log_2(\frac{3}{7}) \approx 0.99$. As a result we obtain $M(0) = H_{\max} - H_0 \approx 0.65$ and $M(1) = H_{\max} - H_1 \approx 2.18$. Thus $M(0) < M(1)$ is true which makes sense as the order of the items at time $t = 0$ is less than at time $t = 1$.

9.4.1 Measuring the Emergence in the AHS

In this section we demonstrate in an example that the AHS behaves emergent. As stated in Sect. 9.1 the AHS is self-organizing, so we only have to measure the value of emergence.

Figure 9.4 shows an example of 20 different tasks which are distributed by the AHS. The output is created by our hormone simulator, a tool for cycle-accurate hormone loop simulation on multiple processing elements. The rectangles mark processing elements while the numbers in the rectangles represent tasks taken by the corresponding processing element. The color of the numbers represents the relation

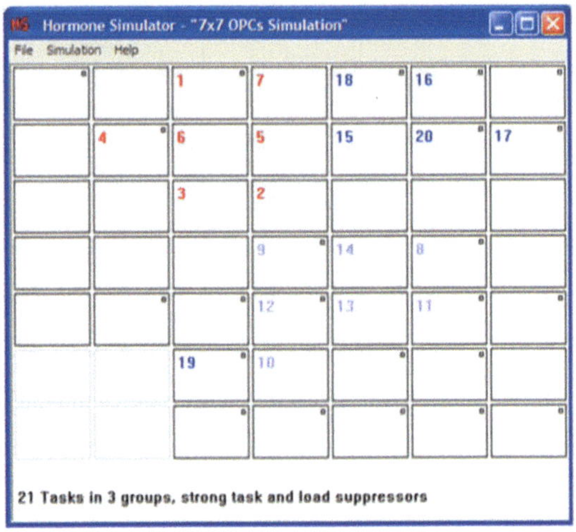

Fig. 9.4 Example for organ clustering

of tasks. Tasks with the same color strongly cooperate and should be placed closely together. In the AHS, this is respected by the accelerator hormones described above attracting related tasks to form so called virtual organs. In the example, the tasks are grouped into three organs, marked red, blue, and light blue. Considering this we are interested in measuring the emergence of each organ using two different strategies.

PE strategy: In the first strategy we measure the emergence as stated in equation (9.2). This means we consider the event "Number of tasks of an organ per PE". The relative frequency p_γ per organ and PE R_γ is then computed by p_γ = Number of tasks of an organ per PE/Overall number of tasks of organ.

Cluster strategy: The second strategy considers the ability of the AHS to cluster related tasks to organs. Tasks of an organ running on neighbored PEs are merged in a cluster C_i. This means we consider the event "Number of tasks in cluster C_i" and compute the relative frequency p_i per cluster C_i and organ by p_i = Number of tasks in cluster C_i/Overall number of tasks of organ. This means we implicitly assume that the tasks of a cluster are running on a single PE, and we set $p_i' := 0$ for the other PEs running tasks of this cluster.

We present evaluations of the emergence of the AHS in the following. We used a benchmark consisting of 64 tasks which are to be distributed in 64 PEs. The tasks are grouped in eight disjunct organs, each of them consisting of eight tasks. The eager values of each PE is initially the same (eager value = 5). In addition, each PE send a strong suppressor as soon as it takes a task (suppressor = 40). Therefore, the AHS allocated each task once at most. If a PE takes a task it will send weak accelerators to the related tasks (accelerator = 2 or 3) in order to favor the clustering of the organs. In addition, each PE can decide each 20 hormone cycles if it offers a taken task for reallocation due to the self-optimization process.

We show the values of the emergence per organ by both the PE and the cluster strategy. The emergence value will be set to -1 if no task of an organ has been taken yet. In general, the number of tasks taken per organ is not constant because of self-configuration and self-optimization. Therefore, we also show the current number of tasks taken per organ in a third diagram. This is necessary to correctly understand and compare the emergence values of different organs.

Figure 9.5 shows the results of the benchmark for 260 hormone cycles. The accelerators are used to build the organs. The values of the emergence of both strategies were periodically computed each 10 hormone cycles. The lowermost diagram in Fig. 9.5 shows that the AHS is able to distribute the tasks of all organs in less than 100 hormone cycles; the time bound for self-configuration given in Sect. 9.3 obviously holds. Organs 6 consists of nine tasks for a short time. The reason is a reallocation of a task because of self-optimization.

Let's consider the values for the emergence now. The maximal entropy is $H_{\max} = \log_2 64 = 6$. This means the maximal emergence value which may occur is $\max M(t) = 6$ for both strategies.

We first consider the results for the cluster strategy. The emergence value for all eight organs reaches 6 at the beginning. However, this is trivial because each organ consists of only a single task at the beginning. In the following the emergence value of the organs 1, 2, 3, 4, 7 and 8 decreases to 4.2 (see organ 4).

The development of the emergence values of organ 6 is different to the others. It shortly drops down to 5.0 in the first 40 hormone cycles and then increases to 6.0. The reason is that this organ becomes a fully connected cluster due to self-optimization and the short-time overallocation of tasks. This also holds for organs 5. The diagram in the middle of Fig. 9.5 shows the emergence measured with the PE strategy. The values of all organs except organ 5 converge to 3. This is because each of these organs needs eight PEs for its tasks ($8 \cdot \frac{1}{8} \cdot \log_2(8) = 3$). Organ 5 needs only seven PEs because of accelerator influence.

We can observe the following from both diagrams: Almost each PE is able to allocate one task at most because of the suppressors. This leads to the emergence value of 3 considering the PE strategy. In addition, the AHS is able to distribute all tasks of the organs 5 and 6 in a fully connected cluster which follows from the emergence values of 6 according to the cluster strategy. Regarding the other organs the AHS is not able to distribute them to fully connected clusters because of the equal distributed eager values for the tasks. Therefore, the tasks of these organs are stochastically distributed to the PEs.

In order to demonstrate the impact of accelerators on organ clustering, we repeated the above experiment with accelerators disabled. The results are shown in Fig. 9.6. The values of the emergence of both strategies were periodically computed each 10 hormone cycles as well as the number of tasks taken per organ.

The AHS needs about 140 hormone cycles to distribute the tasks, see lowermost diagram in Fig. 9.6. The distribution of tasks of organ 5 takes most time because of reallocation of tasks. Here self-optimization interferes with self-configuration due to the short reallocation period of 20 hormone cycles. Therefore, self-optimization starts before self-configuration is finished thus increasing the organ creation time.

Again, the maximal emergence value which may occur is $\max M(t) = 6$ for both strategies.

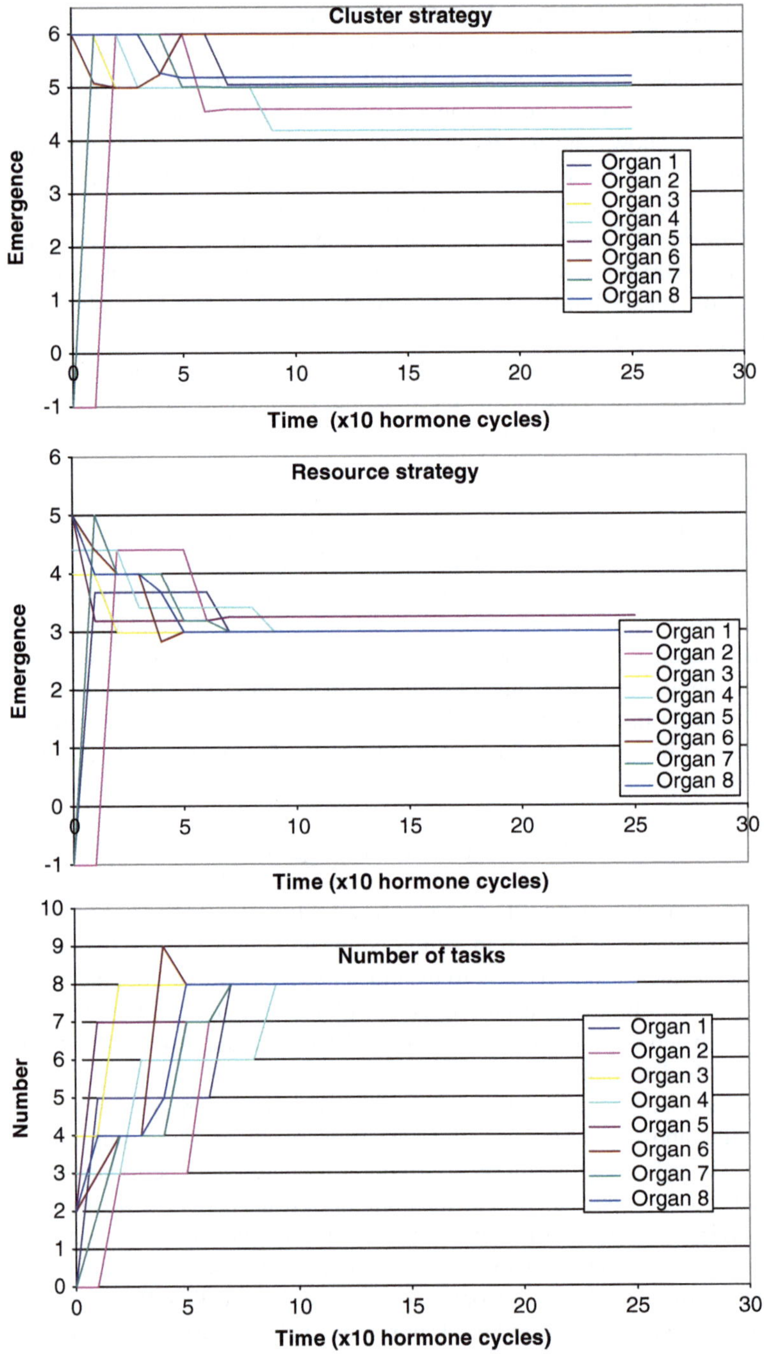

Fig. 9.5 Values of the emergence using the PE and cluster strategy and the number of task allocated per organ. Accelerators are used to favor clustering of organs

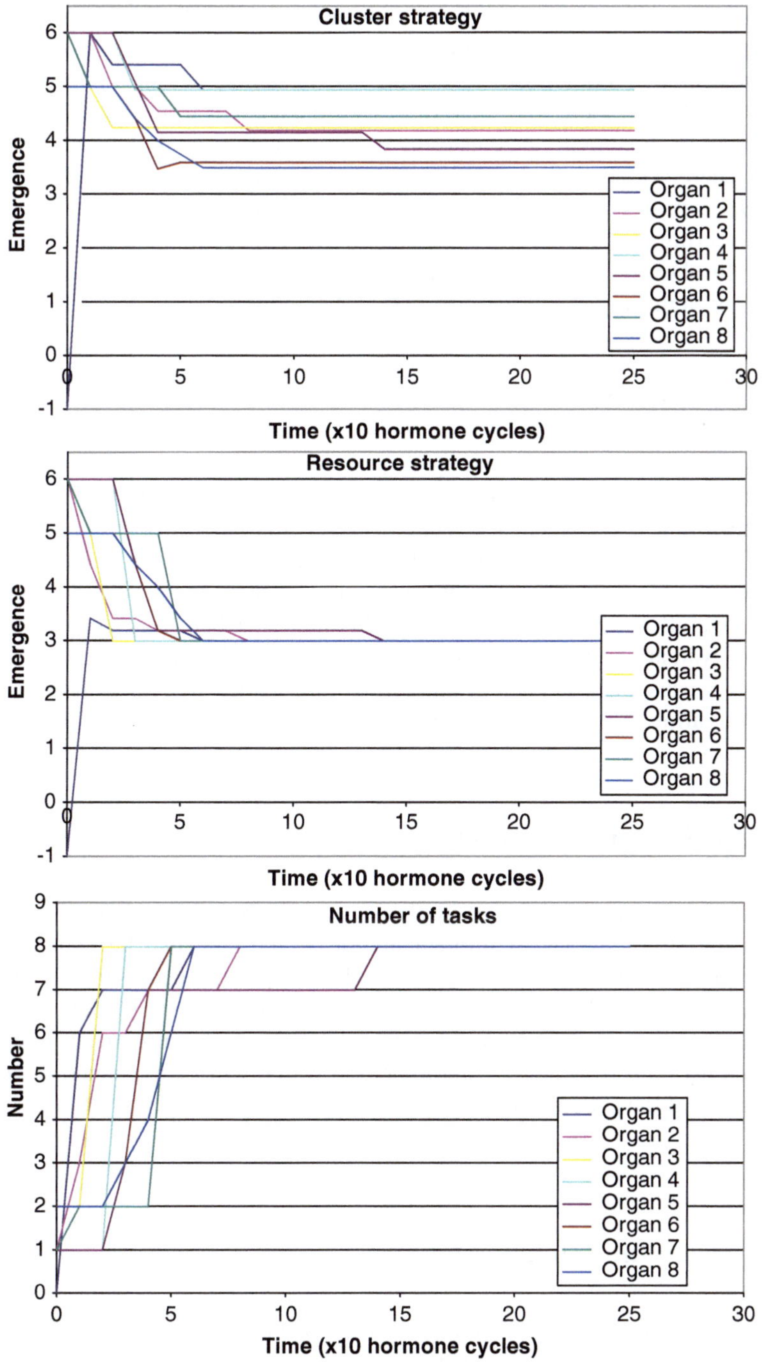

Fig. 9.6 Values of the emergence using the PE and cluster strategy and the number of task allocated per organ. Accelerators are not used

We first consider the results for the cluster strategy. The emergence value for the organs 1 and 4 reach 4.9, the values of the other organs is less. Especially the connection of both organ 5 and organ 6 is moderate shown by emergence values less than 4. Note that their values were excellent in the evaluation before. The reason for this moderate clustering of organs is the equal distribution of eager values. Therefore, the initial task distribution is by random. Because of the missing accelerators this situation cannot be improved by self-optimization.

Considering the PE strategy we obtain (as expected) results similar to the evaluation above. The reason is that each PE may again take one task at most because of high suppressors and the lack of accelerators.

Summary of Sect. 9.4

The cluster strategy evaluates the quality of organ clustering. This can be seen by comparing the results of the first and the second evaluation. It also illustrates the efficiency of the accelerators, especially with regard to their low strength between 2 and 3. The PE strategy evaluates the ability of the PEs to allocate tasks.

However, it is a disadvantage of both strategies that their results depend on the current overall number of tasks taken. This makes it hard to compare results of different evaluations and is the reason why we added a diagram showing the current number of tasks taken to each evaluation. Future work will deal with this problem in order to allow comparisons between measures of different evaluations.

9.5 AHS Implementation

There are several possibilities to implement the AHS. In Sect. 9.5.1, we will first discuss the straight forward implementation which can also be called accurate implementation. We will then (Sect. 9.5.1.1) analyze the necessary memory size and control costs. In Sect. 9.5.2, we will then identify a way to reduce the memory requirements in terms of size and complexity. This newly introduced light-weight approach uses a self-synchronization method which will be explained in detail. A comparison of these two AHS versions in regards to memory size will be shown in Sect. 9.5.3.

9.5.1 Accurate Implementation

The accurate implementation requires a memory for the hormones in which the last received hormone of each hormone source has to be stored. This is due to the fact that the hormone cycles of the single processing elements are not synchronized and

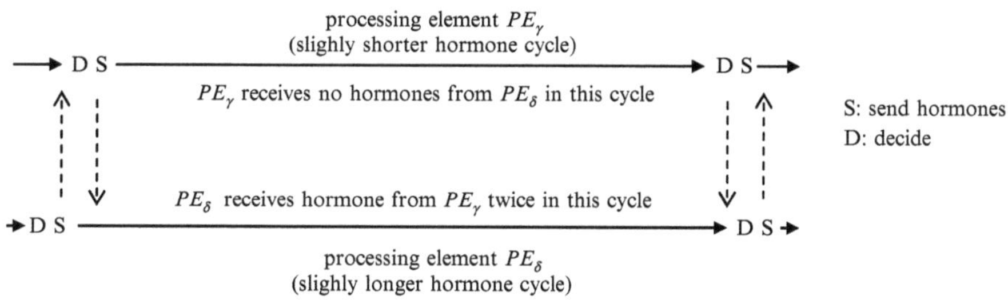

Fig. 9.7 AHS synchronity issues

have only roughly the same cycle duration. Therefore, time shifts can occur which might result in either double or no reception of a specific hormone in this cycle (see Fig. 9.7).

By storing the hormones, this behavior can be encountered. A double occurrence of hormones can be identified and handled by overwriting the previous value, missing hormones can be compensated by falling back to the last received hormones. Out-dated hormones will be identified via an expiration date and will be removed. Therefore, a hormone memory and a hormone memory control unit is necessary to update, remove expired and sum up hormones, as well as to identify the largest eager value. For arithmetic operations, addition, subtraction and comparison operations are needed.

9.5.1.1 Memory Requirements

In Fig. 9.8, the schematic setup of such a hormone memory is displayed for one task. The necessary size of the hormone memory can be calculated as follows:

For each task of a processing element, each last received hormone of each source has to be stored. There are:

- Eager values—“e”: the number of processing elements that apply for the execution of this task
- Suppressors—“s”: the number of processing elements that execute this task
- Accelerators—“a”: the number of neighboring processing elements which execute related tasks multiplied by the number of related tasks

The memory has to be designed for the worst case scenario:

$e \leq \omega$ with ω being the number of processing elements in the system: every processing element could apply for the execution of this task.
$s \leq \omega$: every processing element could be executing the task.
$a \leq \varphi \cdot v_{\max}$ with φ being the number of neighboring processing elements and $v_{\max}$ being the number of related tasks.

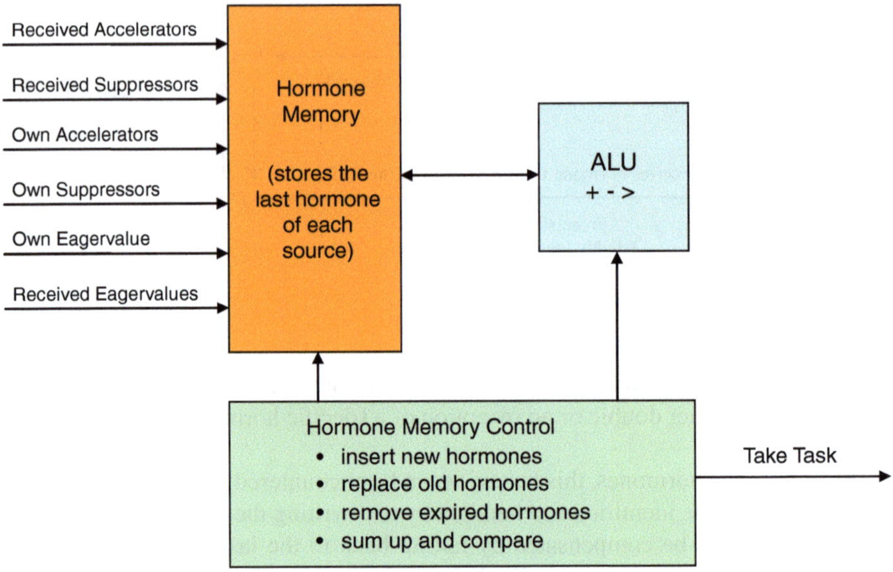

Fig. 9.8 Schematic setup of the hormone memory for one task

Hence, the memory requirements S per task can be calculated as follows:

$$S = e + s + a \leq 2 \cdot \omega + \varphi * v_{\max} = S_{\max}$$

Every entry of the hormone memory has to contain the hormone value, a hormone source identifier (to be able to update the value by a newer value of the same source), as well as a time stamp (for expiration of old values). Let's assume 1 Byte for the hormone value, 2 Bytes for the source address and 4 Bytes for the time stamp—than we have a demand of 7 Bytes per entry.

Hence, the memory requirement results in $7 \cdot S_{\max} \cdot$ Bytes per task.

9.5.2 *A Simplified Light Weight Implementation Using Self-Synchronization*

On powerful processors and microcontrollers the accurate implementation (see Sect. 9.5.1) can be realized without any problems. For simple microcontrollers or direct implementation in hardware, it would be desirable to reduce both the memory requirements and the complexity of the hormone memory controller. In the following, we will present such an alternate approach.

In this alternate approach, the AHS will simply sum up the suppressors and accelerators and determine the highest of the received eager values without storing

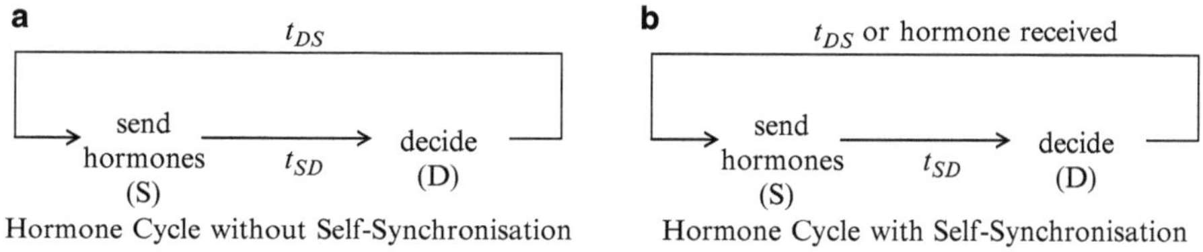

Fig. 9.9 Hormone cycle without and with self-synchronization

individual hormone values. Only the current and the recent sum respectively the current and recent highest eager value will be stored. This change will reduce the costs and complexity dramatically compared to the solution with the hormone memory.

However, this modified setup comes with a problem: the counter measures used by the accurate implementation to compensate the double occurrence or non-occurrence of hormones of the asynchronous working processing elements do not work anymore. This is because the hormones of the different sources are not stored separately any longer and therefore single missing or doubled hormones cannot be compensated.

To solve this problem, a self-synchronization mechanism for the AHS is introduced. Figure 9.9 compares the asynchronous (a) and the self-synchronizing (b) hormone cycle.

Generally, a hormone cycle consists of a hormone sending phase, where all hormones of a processing element are emitted, and a decision making phase to decide on a task to take. The time between sending the hormones and making the decision (t_{SD}) is always fixed (defined in the AHS by a fixed value). This is true for both versions of the hormone cycle. The time between decision making and sending of the hormones is fixed as well for the asynchronous hormone cycle (t_{DS}). In case of the self-synchronizing hormone cycle, this time is set to whichever event comes first: either the waiting time of t_{DS} or the time when the first hormone is received. Using this principle, self-synchronization works as follows:

→ The chronologically first processing element in the hormone cycle will wait the full period of t_{DS} and will afterwards start sending the hormones.
→ The chronologically later processing elements will receive these hormones and will begin their new cycle as well by sending out hormones them selfs.
→ This way, the hormone cycles of the processing elements will synchronize without a central synchronization instance.
→ The AHS has a decentralized synchronization mechanism and therefore it is self-synchronizing.

To achieve a self-synchronizing start-up of all the processing elements' hormone cycles, special care has to be taken as shown in Fig. 9.10.

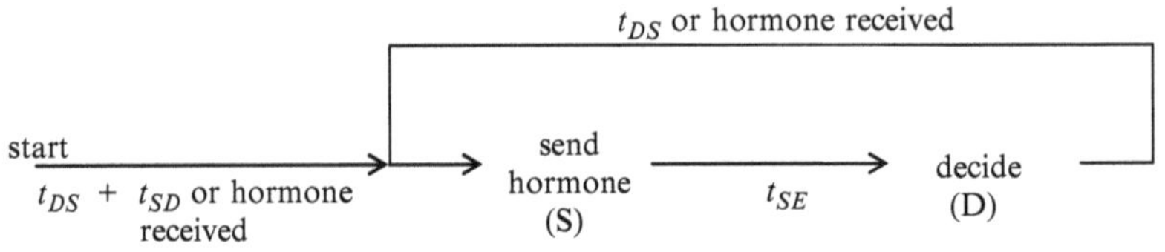

Fig. 9.10 Starting a self-synchronizing hormone cycle

When a processing element starts its cycle, it will wait one entire hormone cycle $t_{DS} + t_{SD}$ before it starts sending out hormones. This will ensure a self-synchronized start-up:

→ If the processing element is the first element starting its cycle, it will wait the full hormone cycle of $t_{DS} + t_{SD}$ and will afterwards start sending out hormones.
→ Processing elements starting later (even additionally introduced elements), will receive hormones of earlier started processing elements during their initial wait period and then start their cycle synchronized to the others.
→ Because a new element might come to life in the middle of a hormone sending phase, it will start its hormone cycle immediately but very probably will have missed some hormones. Therefore, starting processing elements will be sending hormones beginning with the first cycle, but will start their decision process not before the second cycle.
→ All processing elements will start self-synchronized.

9.5.2.1 Memory Requirements

Several considerations have to be made to calculate the memory requirements of the light weight implementation.

While the received hormone values in the current hormone cycle are being accumulated (respectively maximized), the calculated sum of the accelerators and suppressors (respectively the greatest eager value) of the previous cycle has also to be available. The stable values of the previous cycle are necessary to calculate the modified eager values (that has to be sent out) in the current cycle, while the hormones of the current cycle are just being received. Like in a pipeline, in addition to the accumulator register for the current cycle there has to be a cycle register which holds the value of the previous cycle. However, the own hormone values for the calculation of the modified eager value of a task should come from the current and not from the previous hormone cycle.

This has two reasons: First, in contrast to the received values, the self-produced values are already available. Second, a problem would occur if the own hormone values would come from the previous hormone cycle (and not the current). In case of a task allocation, the corresponding suppressors are sent out in the hormone sending phase directly following the allocation of the task. However, the eager values sent in the same phase would not respect these suppressors if own hormones from the

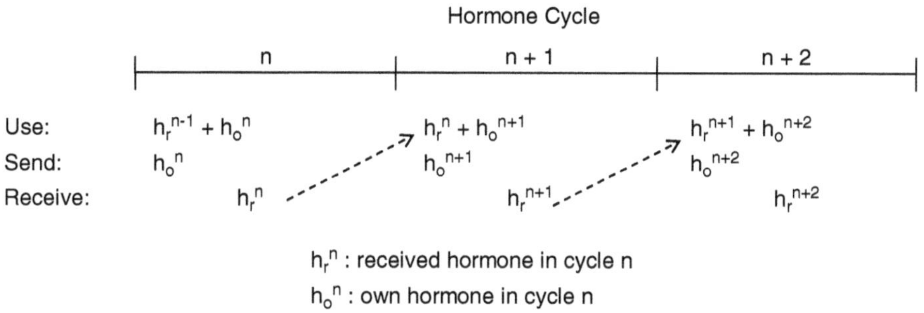

Fig. 9.11 Timing of the light weight implementation of the AHS

previous cycle would be used. In this way, inconsistent hormone values (already suppressors but still high eager values) would be sent out. This inconstancy would lead to a delay of the task allocation.

This effect can be avoided by using the own hormone values of cycle n and the received hormone values from cycle $n-1$ when being in hormone cycle n. This is shown in Fig. 9.11.

Therefore, the own hormones are accumulated separately from the received hormones to be able to get the current values on the one hand and the last values on the other hand. Putting all together, the following structure illustrated in Fig. 9.12 results.

The upper part of this structure accumulates the suppressors and accelerators. While accelerators are added, suppressors are subtracted. There are two accumulator registers, one for the sum of the received accelerators and suppressors, and one for the sum of the own as mentioned above. In contrast to the accurate version with hormone memory, where not only the hormone value but also the hormone source identifier and a timestamp has to be stored, here only the pure values are needed. Assuming 1 Byte as the size of the hormone value, 2 Bytes should be sufficient for the accumulator registers to sum up. Negative values are represented as two's complement. Same is true for the cycle register used to hold the accumulated suppressors and accelerators received in the previous cycle. The lower part of the structure determines the highest eager value. The currently received highest eager value is stored in an accumulator register. Here, the value and the source identifier is stored. The source identifier is used as a second comparison criteria in case of equal highest eager values to achieve a unique decision. So this register has to take 3 Bytes (1 Byte hormone value + 2 Bytes source identifier). Same is true for the following cycle register. The own eager value is fixed for a task. The sum of the received accelerators and suppressors is added to build the modified eager value. The comparison of this modified eager value with the greatest received eager value decides to take the task or not.

With these considerations, the memory requirement per task can be calculated to $3 \cdot 2$ Bytes $+ 2 \cdot 3$ Bytes $= 12$ Bytes.

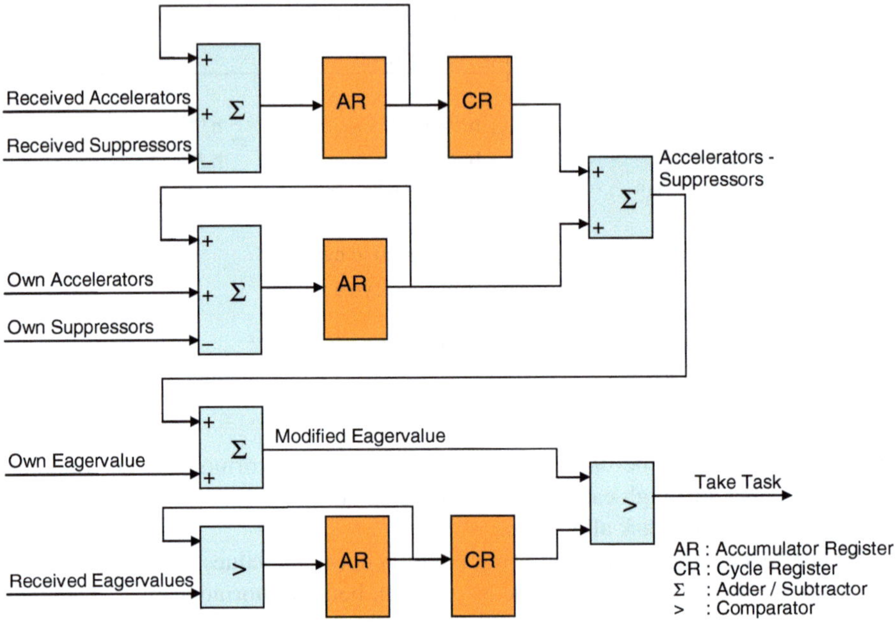

Fig. 9.12 Structure of the light weight implementation of the AHS for one task

So here the memory requirement for a task is a small and fixed value not depending on the number of processing elements or related tasks.

9.5.3 Costs of Implementation

The implementation costs of the light weight version of the AHS are of course considerably smaller. For one, the complex management unit for the hormone memory can be dropped. In addition, the difference in regard to memory requirements per task is:

memory based version:

$$7 \cdot S_{\max} = 7 \cdot (2\omega + \varphi v_{\max}) \text{ Bytes}$$

simplified version:

$$12 \text{ Bytes}$$

Figure 9.13 shows the requirements of the two versions with respect to the number of tasks on a processing element. The dashed lines show the memory requirements of the AHS per processing element with the hormone memory based implementation for two scenarios, a 2×2 processor grid and a 4×4 processor grid.

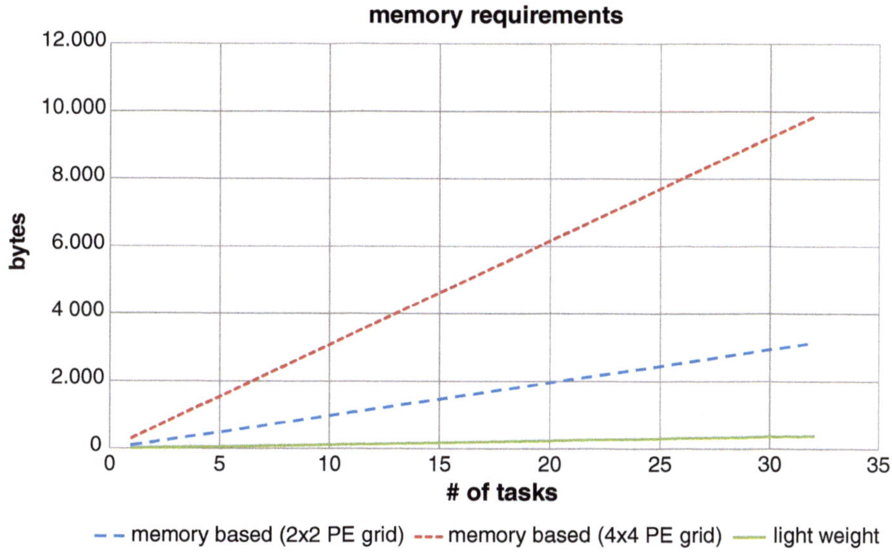

Fig. 9.13 Memory requirements

The continuous line (green) shows the memory requirements per processing element with the light weight version of the AHS. As an example: With the memory based implementation on a 4×4 grid and 16 tasks to map, we need 4,928 Bytes per processing element while the simplified version requires only 192 Bytes.

Summary of Sect. 9.5

In Sect. 9.5, we discussed the hardware requirements—especially concerning memory footprint—of the AHS implementation. To reduce the implementation costs and to be able to deploy the AHS on small microcontrollers and even directly in hardware, we introduced the self-synchronizing version of the AHS. We showed that this self-synchronizing version uses far less memory and eases implementation.

9.6 Application Example

9.6.1 General Concept of MixedCoreSoC

The goal of the MixedCoreSoC project is the research and development of a highly dependable mixed-signal multi-core System-on-Chip. Bio-inspired principles, in particular the Artificial Hormone System presented above, are proposed and

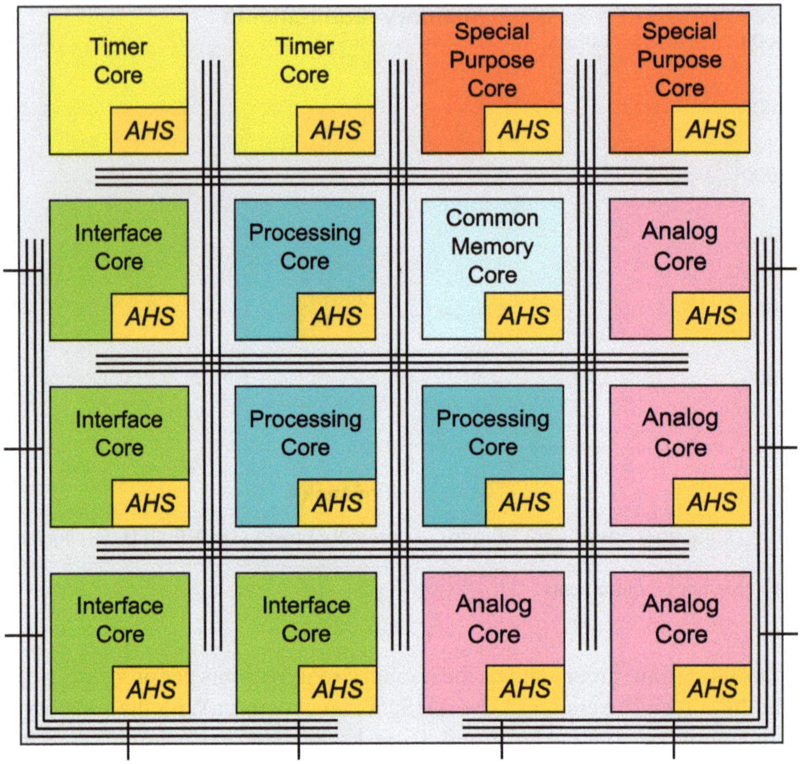

Fig. 9.14 General MixedCoreSoC architecture

investigated to increase the tolerance of the SoC regarding a variety of failures caused by production faults, ageing effects, temperature influence and single event upsets.

The basic idea is to issue a generalized core and task concept for reliable mixed-signal SoC architecture. Therefore, we turned from nowadays likely standard SoC architecture consisting of active digital processing cores and rather reactive digital and analog components like memories, timers, interfaces, converters or amplifiers into a completely heterogeneous many-core architecture, tracing the assumption of a ***generalized core and task concept***. Within this, the entire functionality of the SoC is spread up into different types of cores (Fig. 9.14). In fact, each core type represents a specialized SoC function realized by specific circuits in the domain of analog, digital or mixed-signal. Consequently, each activity of the SoC is considered as a task to be executed on a suitable core. This generalized core and task concept offers a wide range of flexibility by dynamically assigning tasks to cores for proactive and reactive failure handling. To cover a large amount of typical mixed-signal SoC applications, we focuse on the following elementary core templates:

Processing Cores. Widely able to process user specified software tasks, usually coded within any high-level programming language. Comparing to nowadays architectures these cores are equal to general RISC or CISC architectures, exclusively located on digital domain only.

Interface Cores. Interconnection of the SoC to its environment is realized by these cores. Examples are cores implementing parallel or serial links. Conversions among in-out signals and data types, analog/digital conversion, etc. are as well implemented by this instance.

Timer Cores. In terms of time depended task or signal processing (e.g. time constraints etc.), time dependent events are necessary, generated and controlled by instances of this type.

Analog Cores. Processing within analog- or mixed-signal domain is enabled by this type. In general analog cores process complex hardware related tasks such as filtering, amplification etc.

Special Purpose Cores. Reserved black box type for special purposed SoC functions. Necessary application specific circuits of the SoC are represented by this core-type.

Common Memory Cores. Depending on memory architecture of the corresponding SoC, this type implements common memory for global task storage. In fact, this core is also in charge of state- or code-transfer of tasks between cores.

In order to define a correlation between the types of cores and corresponding tasks, the notion of **generalized core** declares the capability of core units, being able to execute one or more generalized tasks. Further the **generalized task** concept covers the classification of all software as well as hardware related tasks. Tasks can be represented as coded programs executable on processing cores or as formal task descriptions defining the activities and timing of interface cores, analog cores, memory cores, etc. To give an example of such a formal task description for other then processing cores, an interface task could be described by the message to send, the sending and receiving task ids, and the requested transmission format (e.g. serial link, asynchronous transmission, baud-rate 19200, no parity, ...). As another example, a timer task could be described by the event id and the period of the event to trigger. Assigning generalized tasks to generalized cores means to execute either the coded program or the formal task description on the selected core. Some cores are able to execute completely different types of task. A timer task can be executed in hardware by a timer core, but as well in software by a processor core. This increases the flexibility and thus the dependability of our approach.

9.6.2 The Role of the AHS

To realize the dynamic assignment of tasks to cores, a middleware architecture is necessary. Here, we have selected the AHS, which is as shown above a highly robust and completely decentralized mechanism to assign tasks to processing elements.

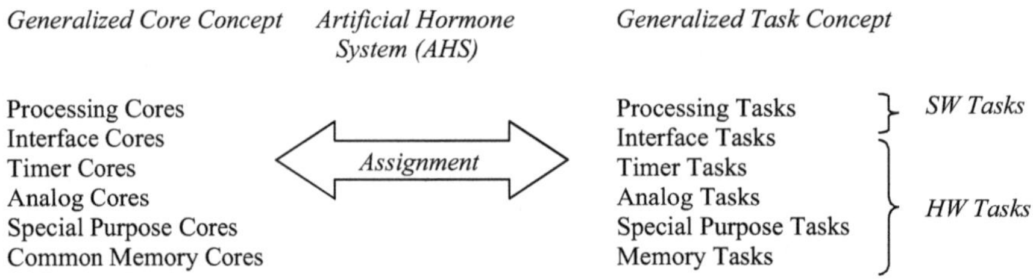

Fig. 9.15 Assignment of generalized tasks to generalized cores by the AHS

The processing elements are represented by the generalized cores of MixedCoreSoC, while the tasks are the geeralized MixedCoreSoC tasks. Figure 9.15 shows the relationship between generalized cores, tasks and the AHS. Task assignment is regulated by balancing hormone levels. Failures induce imbalances in the hormone levels, which automatically lead to task relocation until the balance is reestablished. These hormone balancing control loops allow reactive as well as proactive failure handling.

Figure 9.16 shows an example of the core task assignment, where 3 cooperating tasks to control an ABS braking system have been allocated to corresponding cores. The Wheel Signal Task (a hardware task gathering wheel rounds) has been allocated to a Digital Sensor Interface Core with connection to the wheel sensors. The ABS Brake Task (a software task to avoid blocking of a wheel) has been allocated to a RISC Core. The Brake Control Task (a hardware task to control hydraulic brake pressure) has been allocated to an Analog Valve Driver Core with connection to the hydraulic braking system. If any of these cores fail, the AHS will reallocate the tasks to other suitable cores with the guaranteed real-time bounds as described in Sect. 9.3.

For the MixedCoreSoC project, the AHS had to be adapted to handle not only digital but as well analog cores and tasks. The integration of analog cores lead us to an Artificial Hormone System with Analog components (AHS-A), where digital hormone messages are converted to analog values. As another variant, the Analog Artificial Hormone System (AAHS) maps the principles of the hormone system onto a pure analog hardware architecture. Both will be described in more detail the following.

Digital Artificial Hormone System for Analog Cores

This approach is an adaptation of the digital artificial hormone system to control analog cores (AHS-A). The hormone system is implemented in the simple digital way described above. This means the major hormone loop part of each core is still dealing with digital hormone processing. Therefore, each analog core possesses a local hardware implemented digital hormone loop mechanism (decision stage).

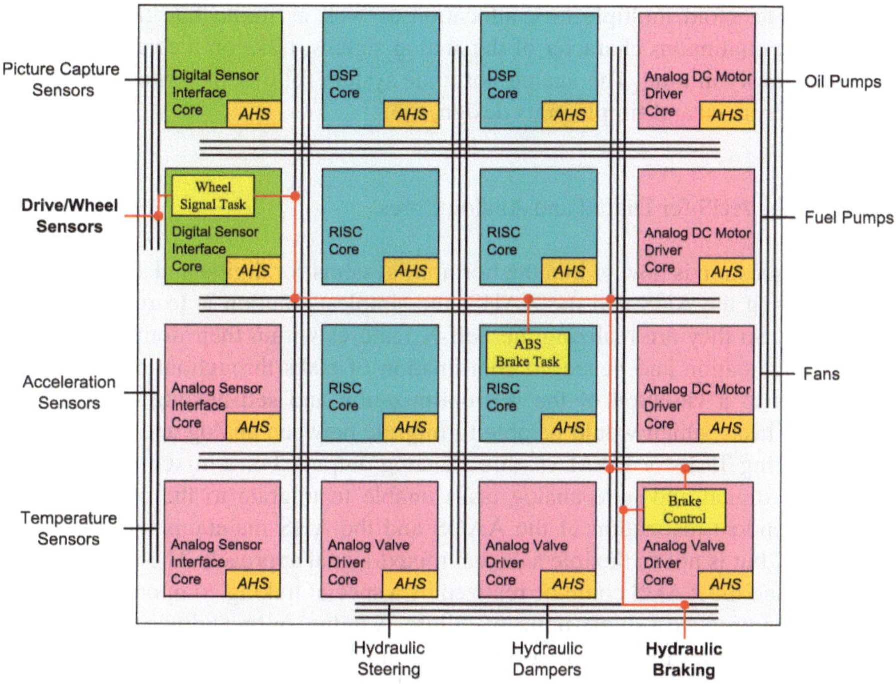

Fig. 9.16 Sample core task assignment to realize an ABS braking component

In order to connect to the analog cores to the loop mechanism as well as the global SoC communication network, analog-digital-converters (ADC) and digital-analog-converters (DAC) are used. In the simplest form, these converters are 1-bit converters to transport the *take task* and *health status* from the digital to the analog domain and vice versa. However, more detailed information about the *health status* can be coded with more bits into the local eager value to improve the task balancing on slightly degraded analog cores.

Analog Artificial Hormone System

Another approach is the concept of the AHS using pure analog components, the analog artificial hormone system, AAHS. The hormone is not a message anymore, which is broadcasted at the network of the SoC, but a signal put on the wires connecting analog cores on the SoC. The hormone level are remodeled as a voltage or a current. In fact, the former iteration steps build by the digital hormone cycles are replaced by continuous time operation. Each core constantly receives signals and is constantly deciding, whether to allocate or drop tasks. According to the digital AHS, whenever tasks are allocated accelerator and suppressor signals are generated to avoid multiple task allocation as well as immediate task drop off. Due to the continuous character of the analog signals, tasks on a core are selected simultaneously in time. The analog hormone system follows the criteria of self-X. It is decentralized and simple in its design, too.

Merging the AHS for Digital and Analog Cores

The next question is how to join the hormone systems for digital and analog cores.

Regarding the AHS and the AAHS, the simplest solution is to run both independently. So they are realizing the Self-X features within their domain, but with no communication and hence no coordination of tasks throughout both systems. This leads to a violation of the self-optimization and self-healing concept with respect to tasks which would be able to migrate between analog and digital cores (e.g. Filtering Tasks or PWM vs. direct analog output). For a mixed-signal design with pure digital and pure analog tasks unable to migrate to the other domain, the independent realization of the AAHS and the AHS maintains the full self-X properties, but is not as flexible as a real mixed-signal approach. The interaction of the AHS and the AAHS could be realized by a special linking component, however this would introduce a single-point-of-failure. A failure or breakdown of the linking component implies the loss of communication between the analog and the digital part. A possible countermeasure is to establish some redundancy here.

Regarding the AHS-A, this problem does not occur. Analog and digital cores are connected by a unique digital hormone system, while hormones are transformed to the analog domain within each analog core. On the downside, this rises the complexity. Furthermore, to allow task migration between analog and digital cores, a sophisticated data switching methodology has to be developed to maintain task communication besides from hormones. If a task migrates from a digital to an analog core or vice versa, it still has to communicate with the other cooperating tasks.

Figure 9.17 shows the already mentioned implementations. On the left AHS is deployed on a simple digital SoC. The right side shows the independent usage of AHS and AAHS on a SoC without hormone interaction. The SoC shown in the middle uses AHS-A. This hormone system is the only one, which connects both, analog and digital cores. Self-X features are met globally and tasks can either be pulled to analog or digital cores, whichever has the highest modified eager value (i.e. for the allocation of a filter task).

MixedCoreSoC is work in progress. A comparison between first implementations of the three approaches is shown in Table 9.1, considering a SoC chip with 10 cores and an occurrence of 5 possible tasks for each core. As can be seen the overhead is relatively low and the performance values are good.

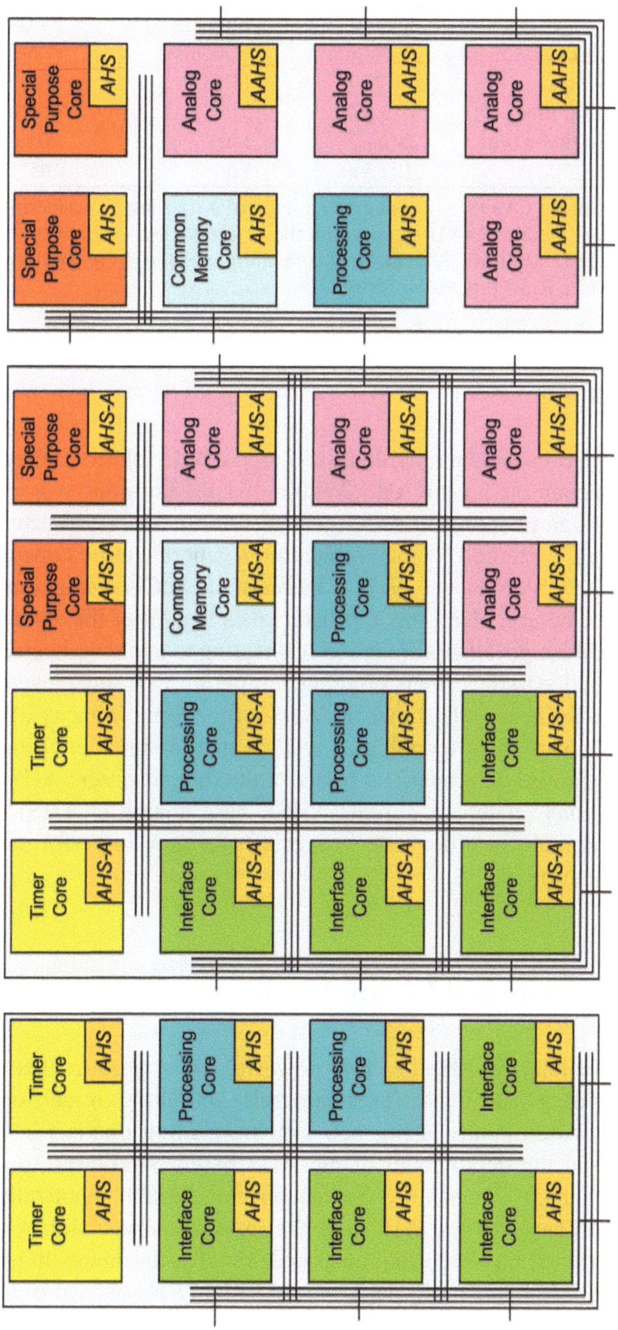

Fig. 9.17 Possible implementations of the hormone system on a SoC (AHS, AHS-A, AAHS)

Table 9.1 Comparison of the hormone system implementations

	AHS	AAHS	AHS-A
Gates/ OP's	23,360	120 OpAmp's	24,040
Needed chip area	0.284mm^2	0.360mm^2	0.293mm^2
Percentage of core	12.75 %	18.75 %	13.1 %
Min. self-config. time	613.5 ns	5 μs	735.75 ns
Min. communication time	20 ns	1 μs	20 ns
Min. cycle time	122.3 ns	5 μs	145.75 ns

Based on a 180 nm CMOS processing technology of Xilinx Inc., considering the total amount of system NAND gates within the design netlist, each given with 2 inputs and at least $12.2\,\mu\text{m}^2$. Percentage of core within AHS, AHS-A considers the use of a MB-Lite RISC processing cores.

Summary of Sect. 9.6

In this Sect. 9.6 we presented a reliable mixed-signal SoC solution as a particularly suited application example for the AHS. Within this, the hormone system is used to control the task assignment on a mixed-signal multi-core SoC architecture. Besides this purpose, which is granted by the architectures generalization concept between cores and tasks, the system is also used to increase the SoC's overall dependability in terms of system failure. Here, the approach takes benefit of the significant AHS self-X properties like e.g. self-optimization or -healing to avoid dedicated cores and task from crashing due to the effects of aging, temperature or single event upset. As another aspect the SoC example shows the AHS's functional capacity of a hormone based task control also in pure analog as well as mixed-signal environment. Last but not least, we showed a comparison between the three different AHS controlled SoC implementations related to digital (AHS), pure analog (AAHS) and mixed signal (AHS-A) domain in order to give a brief overview in performance results and resource consumption.

9.7 Final Remarks and Future Work

As shown in this article, the AHS is a very powerful technique to realize a real-time middleware providing organic features like self-configuration, self-optimization and self-healing. Additionally using self-synchronization enables a lightweight implementation and the deployment of the AHS based middleware on small microcontrollers. Emergent effects are a consequence of the self-organization process. The level of emergence for the AHS and the influence especially of accelerator hormones on this level has been demonstrated. Finally, an interesting application example on a mixed-signal multi-core system-on-chip ha been presented. Here, especially the self-healing capabilities of the AHS are used to increase the system dependability. This is an important issue because shrinking chip feature sizes increases the vulnerability to failures due to aging, temperature or single event upsets.

For the future, we plan to investigate further the behaviour of the AHS. We want to measure the emergence of test scenarios in real-world conditions. Furthermore, self-protection shall be investigated. The self-x features of the AHS increase the system robustness but introduce as well new sources for attacks. Task allocation can be disturbed e.g. by a malicious manipulation of hormone values. Counter measures like immunology shall be investigated here. Finally, a hierarchical version of the AHS is planned to allow handling of very large numbers of processing elements (10000 and more) by the AHS.

References

1. Becker J, Brändle K, Brinkschulte U, Henkel J, Karl W, Köster T, Wenz M, Wörn H (2006) Digital on-demand computing organism for real-time systems. In: Workshop on parallel systems and algorithms (PASA), ARCS 2006. Frankfurt, Germany
2. Brinkschulte U, Pacher M, von Renteln A (2007) Towards an artificial hormone system for self-organizing real-time task allocation. 5th IFIP workshop on software technologies for future embedded & ubiquitous systems (SEUS 2007), Santorini Island, Greece, May 2007, pp 339–347
3. Brinkschulte U, Pacher M, von Renteln A (2008) An artificial hormone system for self-organizing real-time task allocation. Springer, Berlin
4. Brinkschulte U, von Renteln A, Pacher M (2008) Measuring the quality of an artificial hormone system based task mapping. In: ACM autonomics 2008, Turin, Italy
5. Brinkschulte U, von Renteln A (2009) Analyzing the Behavior of an Artificial Hormone System for Task Allocation. In: 6th international conference on autonomic and trusted computing (ATC 2009), Brisbane, Australia
6. Crutchfield JP (1994) The calculi of emergence: Computation, dynamics, and induction. Phys D 75:11–54
7. CSIRO (2009) Centre for complex systems. URL http://www.dar.csiro.au/css/. Accessed on Jan 2012
8. DFG Forschergruppe 1085 (2009) OC-Trust. URL http://www.informatik.uni-augsburg.de/lehrstuehle/swt/se/projects/oc-trust/. Accessed on Jan 2012
9. DFG Schwerpunktprogramm 1183 (2007) Organic computing. URL http://www.aifb.uni-karlsruhe.de/Forschungsgruppen/EffAlg/projekte/oc/inhalte. Accessed on Jan 2012
10. EU (2009) Program Future Emerging Technolgies FET – Complex systems. URL http://cordis.europa.eu/ist/fet/co.htm. Accessed on Jan 2012
11. German Research Foundation priority program 89 (2010) Invasive computing (InvasIC). URL http://invasic.informatik.uni-erlangen.de/en/index.php. Accessed on Jan 2012
12. Holzer R, Meer H, Bettstetter C (2008) On autonomy and emergence in self-organizing systems. In: Proceedings of the 3rd international workshop on self-organizing systems, IWSOS '08. Springer, Berlin, pp 157–169
13. Horn PIR (2001) Autonomic computing manifesto: IBM's perspective on the state of information technology, IBM cooperation
14. IBM (2003) Autonomic computing. URL http://www.research.ibm.com/autonomic/. Accessed on Jan 2012
15. IEEE (2009) Organic computing task force. URL http://www.organic-computing.org/ieeetaskforce/. Accessed on Jan 2012
16. Jetschke G (1989) Mathematik der Selbstorganisation. Harry Deutsch Verlag, Frankfurt

17. Kephart JO, Chess DM (2003) The vision of autonomic computing. IEEE Computer Society, Silver Spring, pp 41–50
18. Lipsa G, Herkersdorf A, Rosenstiel W, Bringmann O, Stechele W (2005) Towards a framework and a design methodology for autonomic SoC. In: 2nd IEEE international conference on autonomic computing, Seattle
19. Mnif M, Müller-Schloer C (2006) Quantitative emergence. In: IEEE Proceedings of the 2006 mountain workshop on adaptive and learning systems (IEEE SMCals 2006)
20. Mösch F, Litza M, Auf AES, Maehle E, Großpietsch KE, Brockmann W (2006) ORCA – towards an organic robotic control. In: Self-organizing systems, 1st international workshop (IWSOS 2006) and 3rd international workshop on new trends in network architectures and services (EuroNGI 2006) Proceedings, LNCS 4124, ISSN 0302-9743, Springer, Berlin, pp 251–253
21. Mueller-Schloer C, von der Malsburg C, Wuertz RP (2004) Organic computing. Informatik Spektrum 27(4):332–336
22. Prothmann H, Rochner F, Tomforde S, Branke J, Müller-Schloer C, Schmeck H (2008) Organic control of traffic lights. In: Autonomic and trusted computing, 5th international conference (ATC 2008). Oslo, Norway, pp 219–233
23. von Renteln A, Brinkschulte U (2007) Reliablity of an artificial hormone system with self-X properties. In: Parallel and distributed computing and systems, Cambridge, Massachusetts
24. von Renteln A, Brinkschulte U, Weiss M (2008) Examining task distribution by an artificial hormone system based middleware. In: ISORC, Orlando, FL
25. Richter U, Mnif M, Branke J, Müller-Schloer C, Schmeck H (2006) Towards a generic observer/controller architecture for organic computing. In: Hochberger C, Liskowsky R (eds) INFORMATIK 2006 – Informatik für Menschen!, volume P-93 of GI-Edition. Lecture notes in informatics (LNI). Köllen Verlag, Germany, pp 112–119
26. Schmeck H (2005) Organic computing – a new vision for distributed embedded systems. In: 8th IEEE international symposium on object-oriented real-time distributed computing (ISORC 2005), Seattle, pp 201–203
27. Shannon C (1948) A mathematical theory of communication. Bell Syst Tech J 27
28. Trumler W, Pietzowski A, Satzger B, Ungerer T (2007) Adaptive self-optimization in distributed dynamic environments. In: 1st IEEE international conference on self-adaptive and self-organizing systems (SASO 2007), Boston
29. VDE/ITG (Hrsg.) (2003) VDE/ITG/GI-Positionspapier Organic Computing: Computer und Systemarchitektur im Jahr 2010. GI, ITG, VDE
30. VDE/ITG/GI (2003) VDE/ITG/GI-Positionspapier Organic Computing: Computer und Systemarchitektur im Jahr 2010
31. Whitaker R (1995) Self-Organization, Autopoiesis, and Enterprises. URL http://www710.univ-lyon1.fr/~jmathon/autopoesis/Main.html. Accessed on Jan 2012
32. Wildermann S, Ziermann T, Teich J (2006) Organic Bus. URL http://www12.informatik.uni-erlangen.de/research/organicbus/?lang=en. Accessed on Jan 2012

MIX
Papier aus verantwortungsvollen Quellen
Paper from responsible sources
FSC® C105338

If you have any concerns about our products,
you can contact us on
ProductSafety@springernature.com

In case Publisher is established outside the EU,
the EU authorized representative is:
Springer Nature Customer Service Center GmbH
Europaplatz 3, 69115 Heidelberg, Germany

Printed by Libri Plureos GmbH
in Hamburg, Germany